Advances in Studies of Aging and Health

Volume 3

This book series is a theme-driven compilation, publishing the latest international research findings by key topics upon roles of social, behavioral, psychological, and contextual factors in associating or determining health and wellbeing of older people, with biomedical components taken into consideration. It applies an internationally comparative perspective and a biopsychosocial framework to understand healthy, active, productive, and successful aging, in order to build up a comprehensive and practical source of knowledge for researchers and practitioners in the field of aging and health.

This book series publishes original studies and tracks new research progresses by the major topics in the field of aging and health; systematically reviews recent research advancements of new concepts, frameworks, and methodology; and frameworks; explores the aging issues through an interdisciplinary perspective, and provides fresh empirical findings and theoretical insights in an internationally comparative manner. The goal is to effectively disseminate new policies and intervention practices on aging, in order for the improvement of quality of life of older persons.

The major themes of this book series include but are not limited to disability, cognitive function, chronic diseases and conditions, happiness and psychological wellbeing, self-rated health, frailty, mortality, living arrangements and intergenerational transfers, retirement and pension reforms, financial health and poverty, elderly abuse and human rights, ageism, long-term care, end-of-life care, aging in place, environmental health, age-friendly cities/communities, social participation, centenarian studies, big data in aging research, resilience and vulnerability, health life expectancies, active aging, successful aging, and healthy longevity.

Li Zhang

Living Longer and Healthier at Older Ages

A Longitudinal Analysis of Chinese Oldest Old's Health and Its Determinants

Springer

Li Zhang
School of Sociology
China University of Political Science and Law
Beijing, China

ISSN 2522-5146 ISSN 2522-5154 (electronic)
Advances in Studies of Aging and Health
ISBN 978-3-031-69772-2 ISBN 978-3-031-69773-9 (eBook)
https://doi.org/10.1007/978-3-031-69773-9

This research was funded by the National Social Science Fund of China (Grant No. 21BRK043, grant recipient: Li Zhang), the Social Science Fund of Beijing Municipal (Grant No. 23SRB008, grant recipient: Li Zhang), and the Distinguished Qian Duansheng Scholar Fund at China University of Political Science and Law. The funders had no role in study design, data collection and analysis, decision to publish, or preparation of the manuscript.

This Springer imprint is published by the registered company Springer Nature Switzerland AG
The registered company address is: Gewerbestrasse 11, 6330 Cham, Switzerland

To my grandmother, parents, and children,
Guilian Liu, Jianguo Zhang, Yanqing Shi
Xiying Lu (Candice) and Xiyuan Lu (Emily)

Preface

Fertility decline has caused rapid population aging worldwide. With no exception, the Chinese population is also aging quickly. According to the seventh Chinese census conducted in November 2020, population aged 0–14 (253 million) and 15–59 (894 million) were about 18.0% and 63.4% of the total population, respectively. There were 264 million people who were 60 and over, which is about 18.7% of the total Chinese population. Among them, 190 million were 65 and over, which is about 13.5% of the total population. As compared to the sixth census conducted in 2010, population aged 0–14, 60 and over, and 65 and over increased by 1.4%, 5.4%, and 4.6%, respectively. Whereas the working age population aged 15–59 decreased by 6.8%. With such a trend, China will eventually face a shortage of working age population and a great deal of challenges due to population aging, including old age support, pension policy reform, family care, medical costs, and others.

One of the effective ways to alleviate challenges caused by population aging is to promote healthy aging in older ages. Under such a proceeding, demographers and sociologists are urged to elucidate factors that may influence population health in older ages and prolong human life span. This book tries to use the newly developed age-period-cohort (APC) approach to explore how a variety of demographic, socioeconomic, and behaviral factors have impacted Chinese oldest old's physical and mental health. It also aims to detect how period and birth cohort have played a role in shaping the oldest old's health outcomes. The manuscript has demonstrated, when a variety of social factors are controlled, how oldest old's health status has changed over time and across birth cohorts. The manuscript makes contributions to the fields of demography, sociology, geriatrics, and so forth.

Looking at the monographs that have been published during the past 15 or so years, most books studying Chinese population aging are policy oriented. This can perhaps be explained by an extremely powerful status of the Chinese government. Among those, a few books published by Springer Publisher have dealt with aging in China in particular. These books are such as *Healthy Longevity in China-Demographic, Socioeconomic, and Psychological Dimensions* edited by Zeng Yi, Dudley L. Poston Jr., Denese Ashbaugh Vlosky, and Danan Gu, Springer Publisher, 2008; *Aging in China: Implications to Social Policy of a Changing Economic State* edited by Sheying Chen & Jason L. Powell, Springer Publisher, 2012; *The Development of Security and Whole Care System for the Aged in China*, edited by

Yanzhong Wang, Springer Publisher, 2018. The book edited by Zeng Yi and associates examined aging issues and health longevity in China from demographic, socioeconomic and psychological dimensions. The book answered questions critical to the aging population worldwide. That is, is the period of disability compressing or expanding with increasing life expectancy and what factors are associated with these trends in the recent decades? Is it possible to realize morbidity compression with a prolongation of the life span in the future? This book is obviously a very important piece that enhances our understanding of several longitudinal survey datasets, including the Chinese Longitudinal Healthy Longevity Survey (CLHLS), as well as healthy longevity in China. The book edited by Sheying Chen discussed topics of bio-medicalization, aging policies, housing, community care, institutional elder care, family care, pensions, and mental health in China. The majority of the chapters addressed how social policies and economic reforms have impacted old age care in China. The book edited by Yanzhong Wang devoted more attention to the development of a pension security and comprehensive care system for elders in China. The book provided a large number of comparative analyses of security systems for the aged in China and in more developed Western countries. In addition, the book analyzed the reforms of Chinese pension security system over the past years. Similar to the book edited by Sheying Chen and colleagues, this book is also largely policy oriented.

This current manuscript differentiates it from the above books in the following aspects: First, in contrast to other existing books which are largely policy oriented, this study focuses on studying Chinese oldest old's health and its determinants at the individual level. It is one of the few books that examine Chinese oldest old's health-related factors. Second, the study goes beyond analyses of cross-sectional datasets to a longitudinal scope when exploring the multivariate relationships between health and a number of factors. The study examines how a variety of covariates have affected Chinese oldest old's health status. More importantly, the research reveals the compound effects of age, period and cohort in the associations between the oldest old's health and diverse demographic, socioeconomic, behavioral, and lifestyle covariates.

The book commences with a review of the existing literature on older adults' health determinants. The study then introduces the age-period-cohort approach and the strengths of using this method to study older adults' health. The book introduces several important surveys that contain longitudinal information of Chinese older adults' health, including the Chinese Longitudinal Healthy Longevity Survey (CLHLS), the one that is used to conduct the analyses in this monograph. The following chapters conduct empirical analyses and present regression results, which show how a variety of demographic, socioeconomic, social, behavioral, and lifestyle factors influence Chinese oldest old's physical and mental status. In addition, the research elucidates how period and birth cohort effects play a role in the associations between Chinese oldest old's health and its determinants. The book offers important implications in terms of how to improve older adults' health status and help them to live longer and healthier at older ages.

Since this book studies aging issues and has important policy implications, the book will be of interest to a range of readers. Scholars of gerontology and geriatrics may find this research cutting edge because this is one of the few frontier studies that systematically examine Chinese oldest old's health from a longitudinal scope. This book can also be used as a textbook in graduate and upper-division undergraduate sociology and demography core courses. Beyond the classroom, the book would also be of significant interest to a general audience, including policy makers who are interested in the topics of old age care, healthy aging, and population policy making.

Beijing, China
April 25, 2024

Li Zhang

Acknowledgments

This book represents several years of my research on health and gerontology at China University of Political Science and Law (CUPL). A number of people have offered their insights, assistance, and support while I worked on this manuscript. Special thanks are owed first to my students Rongxing Gu, Gen Zhang, Simiao Wang, Wanni Yang, Pei Wang, and Xingrui Gao. Rongxing helped me on the methodology and data analyses. Gen Zhang assisted me with literature review. The other four students assisted me with literature search. The book would not have been possible without their help. My heartfelt thanks also go to my family, especially my grandmother, parents, and daughters, for their understanding of my absence and providing support and love throughout my time working on this manuscript. This book is as much theirs as it is mine. Special thanks also go to Drs. Danan Gu and Qiushi Feng, the two editors of the book series *Advances in Studies of Aging and Health*, who provided extremely helpful suggestions to improve the manuscript. Lastly, I want to express my deepest appreciation to Evelien Bakker, Manonmani Pradeepkumar, Arun Santhosh Kannanand, and other editors and personnel of Springer, who have been understanding all the time and helped me along the way and have been a great pleasure to work with.

Contents

The original version of the book has been revised. A correction to this book can be found at https://doi.org/10.1007/978-3-031-69773-9_11

List of Figures

List of Tables

Part I

Introduction

Introduction and Overview 1

Low fertility coupled with a longer life expectancy due to economic development and improvements in medical technology have resulted in rapid population aging worldwide. The Statistical Office of the European Communities, the Population Reference Bureau, and the US National Institute on Aging (NIA) identified nine emerging trends in global aging. These trends are: the overall population is aging; life expectancy is increasing; the number of the oldest old aged 85 and over is rising; non-communicable diseases are becoming a growing burden; some populations will shrink in the next few decades; family structures are changing; patterns of work and retirement are shifting; social insurance systems are evolving; and new economic challenges are emerging (Mollah, 2021). As of now, both developed and developing countries are facing population aging. According to the United Nations statistics, the proportion of older adults aged 65 and over in global population increased to 10.0% in 2023 (United Nations, 2024). The proportion of older adults in more developed nations is even higher. For example, in year 2023, population aged 65 and over accounted for 29.6% of the Japanese total population. The percentages of older adults aged 65 and over were 22.8% in Germany, 24.2% in Italy and 21.7% in France (United Nations, 2024). A common distinction between age groups in late life is the division of older adults into three subgroups: young-old (65–74), old-old (75–84), and the oldest old (85 and over). The United Nations (2024) predicted that the world population of the "oldest old" who are aged 80 and over will triple to 446 million by 2050.

Aging is considered as a story of human success, which demonstrates great improvements in general living conditions, public health care, and medical treatment (Goldman et al., 2018). Nevertheless, "the aging population problem" has led to a global concern for the sustainability of the welfare states. In some developed countries, people are living longer and healthier. Although living longer is a great achievement, a longer life expectancy also brings significant challenges. This is because the aging population is affecting economic growth and many other issues, including economy, migration, disease patterns and prevalence, health care, medical

L. Zhang, *Living Longer and Healthier at Older Ages*, Advances in Studies of Aging and Health 3, https://doi.org/10.1007/978-3-031-69773-9_1

expenditure, pension system, family support, community services and et al. Health related issues also become a major concern since the majority of challenges aging brings to people and the society largely link to health related issues. Health of older adults includes both physical and mental health, which greatly impacts older adults' wellbeing and quality of life. Thus, with the trend of global aging, health related issues among older adults are more prominent than ever.

1.1 Aging and Health Related Issues Worldwide

Chronic diseases, major indicators of physical health, are closed linked to one's age. Although more developed countries and regions have relatively well-established healthcare and medical resources, a high incidence of chronic diseases still occur among older adults in those countries due to population aging. Diseases such as heart disease, stroke, diabetes, chronic respiratory diseases, and cancer are commonly diagnosed chronic diseases among older adults. Approximately 17 million people die from cardiovascular diseases each year, 463 million people die from diabetes, 140 million people suffer from chronic obstructive pulmonary diseases, and about nine million people die from cancer annually, with the majority of patients being older adults (World Health Organization, 2021). Unhealthy lifestyles, such as sedentary life styles, excessive smoking and alcohol consumption, and regular consumption of high-salt, high-sugar, and high-fat foods increase the likelihood of developing chronic diseases among older adults (Daw et al., 2017; Han et al., 2017; Lawrence et al., 2017; Murtaugh et al., 2010).

Economically disadvantaged countries and regions face issues of malnutrition among older adults, including deficiencies in vitamins, inadequate protein intake and lack of essential minerals. These factors contribute to a decline in the immune system and the occurrence of chronic diseases among older adults. Healthcare resources, including medical facilities, healthcare professionals and infrastructure, are quite limited in underdeveloped countries and regions. It prevents older adults from receiving timely treatment and management for diseases, which further deteriorates older adults' health.

Besides physical health, mental health issues are also alarming. Depression and anxiety are more commonly observed among older adults than other age groups. When facing chronic diseases, financial difficulties and feelings of loneliness, older adults may experience excessive worry and fear, which develops anxiety disorder among elders over time. The prevalence of depression among older adults is very high. According to estimates of the World Health Organization (WHO), approximately 300 million older adults worldwide suffer from depression, with the majority of them not receiving professional treatment (World Health Organization, 2017). Meanwhile, due to the loss of loved ones, social isolation and physical health problems, older adults experience significant emotional fluctuations and loss of hope and confidence in life. They may even commit suicide in later life (Jang & Tang, 2016; Li et al., 2015; Tsai et al., 2013; Yang et al., 2021a; Yang, 2007; Zhang et al., 2021; 2022). The rate of suicide among older adults remains high with over one million people taking their own lives worldwide each year. Among those who committed suicide, a high percentage of them were individuals aged 65 and above (World Health Organization, 2017).

High prevalence of physical and mental illnesses among older adults brings a heavy burden to both families and the society. Thus, it becomes essential to study factors that influence older adults' health, investigate solutions for treatment, and provide health care and effective disease management services to older adults. Such practices can reduce the need for hospitalization and emergency visits, which alleviates the burden on healthcare systems.

From a sociological perspective, older adults are social capital. They possess abundant wealth of life experience and knowledge that can provide important resources to the society (Fung & Carstensen, 2019). According to social contract theory, human society is built on a mutually supportive relationship, which requires members of a society to care for and support vulnerable groups (Harvey et al., 2020). It is also essentials to promote equal distribution of health rights to people of all age groups. Older adults should be ensured to be able to access the same medical resources and health services as other groups without experiencing unfair treatment due to an old age (Gupta et al., 2020).

Indeed, a variety of actions have been taken to advance older adults' health condition among world countries in recent decades. In 1993, Japan enacted the "Basic Act for a Long-Term Society," aiming to address the challenges of aging population and promote health and well-being of older adults (Japanese Cabinet, 1993). In 2010, the European Union introduced the "European Charter of the Rights of Older People", which outlined the basic rights that older people are entitled to, including health, social participation, education, culture, social security, and called on its member countries to advance health of older adults (European Commission, 2010). Starting from 2016, WHO launched the "Global Strategy and Action Plan on Aging and Health 2016–2020". It advocated policies and practices targeting towards older adults. The plan primarily focused on promoting healthy aging, improving healthcare and social welfare for older adults, and encouraging their participation in social and economic activities. China has also introduced the "Healthy China 2030 Plan" which emphasized that health is an essential requirement regarding human development and argued that health is a fundamental condition for economic and social progress (Xinhua News Agency, 2016). Healthy aging involves engaging in good quality of life and independent living. How to live not only longer but healthier in old ages and ease health care burden and minimize medical expenditure have become hot topics in various fields of world countries.

1.2 Aging and Health Issues in China

China, a country that has practiced the one-child policy for 34 years, is also experiencing low fertility and population aging with no exception. In 2010, the proportion of aging population in China was 8.1%, it rose to 8.7% in 2013, 9.8% in 2016, 11.5% in 2019 and 13.5% in 2021 (National Bureau of Statistics of China, 2022).

Chinese population is aging rapidly and China can be considered as a demographically developed but economically developing country. As of now, China has completed the epidemiological transition from the stage that deaths are caused primarily by infectious diseases to the stage that deaths are caused mainly by chronic diseases. For the Chinese population, the top ranked leading causes of death in year 2021 were heart disease, cardiovascular disease, tumor, respiratory disease, injury, endocrine, nutritional and metabolic disease, digestive system disease, nervous system disease, urogenital disease, and infectious disease (National Health Association of China, 2022). Based on newly released data, in year 2019, about 2.3 million older adults died because of cardiovascular disease, 3.0 million of them died from cancer, those who died from respiratory disease and diabetes were 0.9 million and 0.4 million, respectively (National Health Association of China, 2020). With the increase of aging population, old age care and medical expenditure will continue to be serious challenges to China and its population.

According to Chinese culture, longevity, happiness, wealth, good fortune, and prosperity are considered as five fortunes for thousands of years. Longevity has long been a pursuit of Chinese population. Culture wise, Chinese have been heavily influenced by traditional filial piety ethics for thousands of years. Thus, when older parents' health condition deteriorates, adult children often choose to put aside their work or switch to jobs that make them more convenient to take care of elder parents. This has long been a practice that carries significant meaning for passing on the filial piety tradition in China. With an increasing trend of aging population, the need for old age care rises dramatically. When a higher percentage of adult children or hired caregivers are involved in old age care, a shortage of labor supply in the overall labor market may occur, which will result in an imbalance of supply and demand in the Chinese labor market. The shortage of labor supply and rising costs of labor will eventually limit economic growth of China in a long run. A decrease in labor supply will also reduce the number of people who pay for social security taxes and fees. Together with a rising number of senior population, an increase in social security expenses will occur. It will eventually add financial pressure to the country. Therefore, improving health status of older adults is beneficial, which prolongs the duration their stay in the labor force. Only with more people entering and staying in the workforce and contributing to social security funds, the funds in the national social security account will be sustainable. Considering China is still an economically developing country, how to use available resources to prevent chronic diseases and launch effective disease prevention strategies to promote national health, especially health of older adults becomes an urgent task in the years to come.

1.3 Existing Studies of Older Adults' Health

Since the late twentieth century, a considerable amount of attention has been paid to the study of older adults' health from multiple disciplines. There has been a growing body of literature focusing on such a topic. In the field of biology, research on older adults' health has primarily focused on studying their physiological functions.

Through examining factors such as blood, urine, and DNA, researchers evaluated the immune function, inflammation levels, and genetic diseases of older adults to detect how these factors have affected their physiological health (Fülöp et al., 2017). Clinical and experimental research in medicine has applied strategies such as observing and documenting clinical treatments and disease progression among older adults. Randomized controlled trials have also been used to assess the effectiveness of interventions in curing illnesses among older adults (Bartlett et al., 2017). Epidemiological studies have further investigated the occurrence, distribution of diseases in various populations. Thus far, this line of studies has conducted cross-sectional surveys, longitudinal studies and cohort analyses to evaluate older adults' health status and the associated risk factors (Wang et al., 2022). Psychology has focused on studying the psychological state, cognitive functions, and mental health issues of older adults. Using methods such as cognitive assessments, psychological questionnaires and brain imaging strategies, researchers have gained insights into cognitive function, psychological adaptation, and effectiveness of psychological interventions when studying health of older adults.

Sociologists, on the other hand, have concentrated on evaluating social environment, social support, family caregiving, intergenerational relations, social participation, social policies, arts, media, culture, medicine, migration, pension, employment, and lifestyles of older adults and linked those factors to individual health (Weil & Cabrera, 2021). Methods such as social surveys, social network analysis, and lifestyle assessments have been used to understand how a variety of factors, including social security, social interaction and social integration can explain the health disparities among older adults (Chai & Mei, 2022; Lee et al., 2013; Luo et al., 2020; Phulkerd et al., 2020; Rizzuto et al., 2019; Zhang, 2021).

In terms of methodology, the commonly used methods in scientific studies of older adults' health are such as experimental method, content analysis, survey method and field studies. Both cross-sectional and longitudinal scopes have been applied in existing studies. When conducting experimental research, researchers set up experimental and control groups. Through observing differences between these two groups, scholars were able to determine the effectiveness of the interventions or treatments in improving older adults' health. Researchers who utilized the content analysis approach have reviewed and analyzes literature on older adults' health related issues. The survey methodology has been widely used in empirical studies of health inequalities among older adults, which yielded abundant literature in the fields of sociology and demography (Buchman et al., 2012; Hofer et al., 2019; Huxhold et al., 2020; Inoue et al., 2020; Joe & Demiris, 2013; Kelsey et al., 2010; Khalaila & Vitman-Schorr, 2018; Kim et al., 2017; McDonnall, 2011; Rizzuto et al., 2019; Weaver & Roberto, 2017).

As the country with most aging population in the world, China has received a considerable amount of research attention in recent decades. In the fields of sociology and demography, scholars have conducted abundant studies examining Chinese older adults' health and its related issues, including frailty, morbidity, mortality, risk factors and health determinants. The commonly used databases are such as the China Family Panel Studies (CFPS), the China Health and Retirement

Longitudinal Study (CHARLS) and the Chinese Longitudinal Healthy Longevity Survey (CLHLS).

A range of topics have been covered by these studies and scholars have showed how these factors link to older adults' health. These topics can be categorized as follows: (1) ***health lifestyles*** and its association with older adults' health (Martín-María et al., 2020; Martinez-Gomez et al., 2017; Zhang, 2021); (2) ***intergenerational relations, social support*** and older adults' health (Cong & Silverstein, 2008; Han et al., 2017; Hayotte & Brunson, 2022; Jiang & Zheng, 2013; Li et al., 2006; Li & Zhang, 2023; Liu, 2014; Zhang & Chen, 2014; Zhang & Silverstein, 2022); (3) ***living arrangements*** and older adults' health gaps (Fenga et al., 2017; Lee et al., 2020; Samanta et al., 2015; Wei et al., 2022; Yang et al., 2021b); (4) ***internet use*** and its link to older adults' health outcomes (Huang & Zhang, 2017; Yang et al., 2021a; Yuan, 2021; Zhang, 2016; Zhang & Li, 2022; Zhang & Zhang, 2021); (5) ***grandparental child care*** and grandparents' physical and mental health status (Chen et al., 2011; Guo et al., 2008; He et al., 2021; Xu et al., 2012; Zhang et al., 2015; Zhou et al., 2017); (6) ***social activities and participation*** and their impacts on older adults' health condition (Han et al., 2017; Li et al., 2013); (7) ***disability, functional decline and mortality*** among older adults (Chen et al., 2015; Lv et al., 2019; Zhao et al., 2017); (8) the availability of ***health care/community services*** and its influence on older adults' health (Li et al., 2015; Zhang, 2004; Zhou & Walker, 2016).

It worth mentioning that the majority of existing studies on Chinese older adults' health are cross-sectional in nature. Only very limited research has applied a longitudinal scope to shed light on the changing patterns of Chinese older adults' health and its determinants. Important themes that have been covered by existing longitudinal studies are such as trajectories of frailty among older people in Hong Kong and mainland China (Li et al., 2022; Yu et al., 2018), trends of cancer incidence among Chinese older adults from 2005 to 2016 (Guo et al., 2022); patterns of HIV or AIDS mortality among Chinese older adults from 1990 to 2019 (Ren et al., 2022); trends in cognitive function among Chinese older adults (Hou et al., 2018; Hu et al., 2021); and effects of long-term health lifestyles on health (Aihemaitijiang et al., 2022).

Even with these analyses on older adults' health, we still know relatively little beyond the health disparities that correlate with the above factors at a static point in time. Thus far, few simultaneous assessments of the age, time period, and birth cohort effects are found in the analyses of health differentials among Chinese older adults. Therefore, the aim of this book is to use the age-period-cohort (APC) approach to examine how Chinese oldest old's health has changed over time and across birth cohorts. The research also intends to explore how health trajectories are shaped by a variety of factors across different birth cohorts, and how historical events may leave imprints on cohorts' life experiences so that diverse health outcomes have occurred. This is one of the first studies that consider the ever changing social contexts and bring a longitudinal view to the research on Chinese oldest old's health.

1.4 Prior Research on the Oldest Old

Commonly, the oldest old refers to people who are aged 85 or older, especially in developed countries (National Research Council & Committee on Population, 2001). As population aging rapidly worldwide, the oldest old population has received remarkable research attention in recent years. Even with abundant literature on the oldest old, research on such a group is still relatively underdeveloped, and this age group remains understudied as compared to their younger counterparts (Wu & Gu, 2021).

Starting from the second half of the twentieth century, especially after the 1970s, life expectancy increased substantially due to social, economic and medical development. Oldest old mortality started to decline in industrialized countries and the oldest old population in these countries began to increase considerably after the 1980s. Based on data released by the United Nations (2024), only about 0.2% of the world's population were aged 85 and above in 1950, such a percentage increased to 0.9% in 2023. Globally, all regions and areas are experiencing population aging and will continue to do so over the next several decades. A more advanced stage of population aging will occur in developed countries, with the proportion of older persons aged 65 and over rising from 20.4% in 2023 to 27.4% in 2050.The growth of the oldest old population aged 85 and above had a greater magnitude in more developed regions where the percentage rose from 0.4% in 1950, 0.7% in 1980, to 2.7% in 2023. The oldest old have become the world's fastest-growing age group.

When it comes to health related issues, the overall physical health status of the oldest old has been found to get better in both developed and less developed countries (Gu et al., 2015). Even with the improvement in physical health over time, disability and dependency are highly prevalent among the oldest old. The disability rate of the oldest old also rises rapidly with age (Gu et al., 2015). Prior research indicated that on average, one-third of the oldest old (ages 80 and above) have ADL disabilities (Di Bari & Balzi, 2017). The ADL and IADL disability rate among centenarians may reach 90.0% (Zeng et al., 2017). Among the oldest old, women showed higher ADL and IADL impairment levels and worse functional capacities although they have a longer life expectancy (Solé-Auró & Crimmins, 2013). Gu et al. (2015) found that disability trends not only differed between young old and the oldest old, between men and women, but also differed within a country and across countries.

Prior research also paid attention to cognitive function decline of the oldest old (Zeng et al., 2017). Linear declines have been found in most of the cognitive domains, including perceptual speed, episodic memory, verbal fluency, vocabulary and inductive reasoning (Ghisletta et al., 2006). A study of the U.S. showed that only 3.0% of individuals who were aged 65–74 had Alzheimer's disease (AD), but the proportion increased to 17.0% among those who were aged 75–84, and 32.0% among the oldest old (Hebert et al., 2013). A study reported that the annual incidence of AD among individuals who were aged 65–74 was 0.4%, it increased to 3.2% among those were aged 75–84 and 7.6% among the oldest old (Rajan et al., 2019). Recent studies also extended the research scope to changing patterns of the oldest old's cognitive function. To illustrate, through analyzing data from CLHLS, a longitudinal study indicated that the cognitive function of Chinese oldest old's has been improving among more recent birth cohorts (Wu & Gu, 2021).

Regarding mental health of the oldest old, studies have found that as compared to young old, the oldest old were more likely to have a higher level of depression (Wu & Gu, 2021). Some scholars reported that in areas such as life satisfaction and loneliness, no significant decline was observed among the young-old group, but a considerable decline was found among the oldest old. The oldest old population expressed lower life satisfaction and experienced a sharp decline in positive affect (Baltes & Smith, 2003). Others, however, have shown that the psychological well-being of the oldest old may not decline with age. Many oldest old are able to maintain a high level of positive affect even with high levels of disability and dependency (Ng et al., 2017). By analyzing data from CLHLS, Yi and Vaupel (2002) found that Chinese oldest old's level of satisfaction towards life did not change much with an increasing age. A study of the U.S. sample showed similar results that the oldest old were able to maintain relatively high levels of well-being (Smith et al., 2008).

Scholars have proposed two pathways to explain the influence of age on oldest old's psychological health. That is, age may negatively influence the oldest old's psychological health through physical and cognitive dysfunction; meanwhile, age can positively affect the oldest old's psychological health through increased ability of gerotranscendence (i.e., adaptation) (Gondo et al., 2013). More empirical studies seemed to support the second pathway by showing that the positive age effect of adaptation surpassed the negative age effect of dysfunction, which led to a positive psychological health outcome among the oldest old (Zeng et al., 2017). Factors such as personality traits, genes, relative income and social resources have been found to be related to the oldest old's positive affect (Etxeberria & Urdaneta, 2019; Gu et al., 2019).

Chronic diseases are highly related to age. Cardiovascular diseases (CVD) and cancers are the most common causes of death among the oldest old. In the USA, based on data from the National Center for Health Statistics (NCHS) (2018), among people aged 80 and older, 335,780 deaths (27.2% of the total deaths) were due to heart disease and 177,715 (14.4% of the total deaths) died from malignant neoplasms (Wu & Gu, 2021). As compared to young old aged 60–80, a significant increase in the prevalence of almost all types of chronic diseases can be observed among the oldest old aged 80 and above.

Poverty is often associated with an old age. The oldest old tend to have a higher risk of falling into poverty as compared to relatively younger age groups. As compared to younger older adults, the oldest old are more likely to spent their savings, to have higher expenditures for medical care and other healthcare services. Based on data released by the Organization for Economic Co-operation and Development (OECD) (OECD, 2019), the poverty rate in OECD countries were 11.6% and 16.2% among people aged 66–74 and 75 and above, respectively.

The oldest old is also the group that demands a high level of long-term care (LTC) due to the high prevalence of comorbidity, disability and widowhood (Wu & Gu, 2021). It was projected that the number of oldest old (aged 80+) in the USA who need LTC would increase by 2.7 times from 2010 to 2050 as compared to

1.7 times for young old aged 65–79 (Zeng et al., 2014). Zeng and associates (2014) also estimated that the number of the oldest old needing LTC would exceed that of young old after year 2030. In China, about 20 million older adults aged 65 and over need LTC in 2020, among which seven million are the oldest old. It was projected that the oldest old who need LTC would increase by about 1.5 times and 4 times by 2030 and 2050, respectively (Wu & Gu, 2021). An increasing demand for LTC will certainly impose great challenges in care provision, especially for countries that do not have sufficient care and economic resources. Since the majority of the oldest old in China live in families to seek support, only about 3.0% of the oldest old live in institutions, Chinese families will face great challenges in LTC in future. Considering a high level of morbidity and disability and a high level of demand for LTC among the oldest old, research focusing on such a vulnerable group will certainly add value for better addressing aging challenges.

1.5 Objectives and Structure of the Book

The definition of the oldest old in this manuscript follows the common definition of the oldest old. It refers to individuals aged 85 and over. The study has two main goals. First, it aims to demonstrate how Chinese oldest old's health status has changed by period and across birth cohorts during 1998–2018. Second, the study intends to investigate how period and birth cohorts have played a role in the associations between Chinese oldest old's health and a variety of demographic, socioeconomic, and lifestyle factors. To achieve the main goals of this study, this manuscript carries out several chapters of empirical studies, touching on demographic, socioeconomic as well as healthy lifestyle factors that have potential influence on the oldest old's health. Chapters in this book are organized in two parts. Part I includes this Introduction and Overview chapter and another two chapters (Chaps. 2 and 3). Chapter 2 reviews literature on health of older adults (including the oldest old), and Chap. 3 introduces data and methods. Specifically, Chap. 3 introduces the period and cohort effects. It elaborates the age-period-cohort (APC) approach and the cross-classified random-effects models (CCREMs) and it highlights the advantages of using CCREMs when studying health of the oldest old. Part II (Chaps. 4 through 9) moves to investigation of Chinese oldest old's health by analyzing empirical data at the individual level. Data are from the Chinese Longitudinal Healthy Longevity Survey (CLHLS) which is highly interested in surveying the oldest old in China. Chapter 4 first describes the changing patterns of Chinese oldest old's health status by time period and across birth cohorts during 1998–2018. The chapter then uses CCREMs to show period and cohort variations when only including age and excluding other covariates in regression models to predict the oldest old's health. Chapters 5, 6, 7, 8, and 9 shift research interests to the oldest old's health determinants. Health determinants studied in separate chapters in this manuscript are consistent with factors that have been found influential on health of older adults (including the oldest old) in previous studies. Chapters 5 and 6 examine the associations between demographic and socioeconomic factors and the oldest old's health, respectively.

Chapters 7, 8, and 9 investigate how physical activities, leisure activities and health lifestyles (i.e. smoking and drinking) have affected Chinese oldest old's health outcomes. Chapter 10 concludes findings and tries to bring together several areas of the book to elucidate how this manuscript improves our understanding of the oldest old's health from a longitudinal point of view. The ending remarks further place future research attention to longitudinal studies of the oldest old's health.

References

Aihemaitijiang, S., Zhang, L., Ye, C., Halimulati, M., Huang, X., Wang, R., & Zhang, Z. (2022). Long-term high dietary diversity maintains good physical function in Chinese elderly: A cohort study based on CLHLS from 2011 to 2018. *Nutrients, 14*(9), 1730.

Baltes, P. B., & Smith, J. (2003). New frontiers in the future of aging: From successful aging of the young old to the dilemmas of the fourth age. *Gerontology, 49*(2), 123–135. https://doi.org/10.1159/000067946

Bartlett, D., Firth, C., Phillips, A., Moss, P., Baylis, D., & Syddall, H. (2017). The age-related increase in low-grade systemic inflammation (Inflammaging) is not driven by cytomegalovirus infection. *Aging Cell, 16*(5), 784–791.

Buchman, A., Boyle, P., Yu, L., Shah, R., Wilson, R., & Bennett, D. (2012). Total daily physical activity and the risk of AD and cognitive decline in older adults. *Neurology, 78*(17), 1323–1329.

Chai, X., & Mei, J. (2022). Investigating food insecurity, health lifestyles, and self-rated health of older Canadians living alone. *BMC Public Health, 22*, 2264.

Chen, F., Liu, G., & Mair, C. A. (2011). Intergenerational ties in context: Grandparents caring for grandchildren in China. *Social Forces, 90*, 571–594.

Chen, W., Fang, Y., Mao, F., Hao, S., Chen, J., Yuan, M., & Hong, Y. A. (2015). Assessment of disability among the elderly in Xiamen of China: A representative sample survey of 14,292 older adults. *PLoS One, 10*, e0131014.

Cong, Z., & Silverstein, M. (2008). Intergenerational time-for-money exchanges in rural China: Does reciprocity reduce depressive symptoms of older grandparents? *Research in Human Development, 5*, 6–25.

Daw, J., Margolis, R., & Wright, L. (2017). Emerging adulthood, emerging health lifestyles: Sociodemographic determinants of trajectories of smoking, binge drinking, obesity, and sedentary behavior. *Journal of Health and Social Behavior, 58*, 181–197.

Di Bari, M., & Balzi, D. (2017). Cardiovascular disease in the oldest old: A geriatric epidemiology perspective Cham. In A. Ungar & N. Marchionni (Eds.), *Cardiac management in the Frail Elderly patient and the oldest old*. Springer Publisher. https://doi.org/10.1007/978-3-319-43468-1_2

Etxeberria, I., & Urdaneta, E. (2019). Subjective well-being among the oldest old: The role of personality traits. *Personality and Individual Differences, 146*, 209–216. https://doi.org/10.1016/j.paid.2018.04.042

European Commission. (2010). *European year for active ageing and solidarity between generations 2012-European pact for mental health and well-being: Declaration on healthy ageing-old age is not a disease*.

Fenga, Z., Falkinghamb, J., Liuc, X., & Vlachantoni, A. (2017). Changes in living arrangements and mortality among older people in China. *SSM – Population Health, 3*, 9–19.

Fülöp, T., Larbi, A., Dupuis, G., & Pawelec, G. (2017). Ageing, frailty, and age-related diseases. *Biogerontology, 18*(6), 721–723.

Fung, H., & Carstensen, L. (2019). Harnessing wisdom: Age-related changes in the use of social resources. *Perspectives on Psychological Science, 14*(3), 400–419.

Ghisletta, P., McArdle, J. J., & Lindenberger, U. (2006). Longitudinal cognition-survival relations in old and very old age: 13-year data from the berlin aging study. *European Psychologist, 11*(3), 204–223. https://doi.org/10.1027/1016-9040.11.3.204

Goldman, D., Chen, C., Zissimopoulos, J., & Rowe, J. W. (2018). Measuring how countries adapt to societal aging. *PNAS, 115*(3), 435–437. https://doi.org/10.1073/pnas.1720899115

Gondo, Y., Nakagawa, T., & Masui, Y. (2013). A new concept of successful aging in the oldest-old: Development of Gerotranscendence and its influence on the psychological Well-being. *Annual Review of Gerontology & Geriatrics, 33*(1), 109–132. https://doi.org/libproxy2.usc.edu/10.1891/0198-8794.33.109

Gu, D., Gomez-Redondo, R., & Dupre, M. E. (2015). Studying disability trends in aging populations. *Journal of Cross-Cultural Gerontology*, 21–49. https://doi.org/10.1007/s10823-014-9245-6

Gu, L., Cheng, Y., Phillips, D. R., & Rosenberg, M. (2019). Understanding the wellbeing of the oldest-old in China: A study of socio-economic and geographical variations based on CLHLS data. *International Journal of Environmental Research and Public Health, 16*(4), 601. https://doi.org/10.3390/ijerph16040601

Guo, B., Pickard, J., & Huang, J. (2008). A cultural perspective on health outcomes of caregiving grandparents. *Journal of Intergenerational Relationships, 5*, 25–40.

Guo, H., Lin, K., Yang, K., Ma, Z., Cao, M., Hu, Y., & Yan, Y. (2022). Trends of cancer incidence among Chinese older adults from 2005 to 2016: A log-linear regression and age-period-cohort analysis. *Frontiers in Public Health*, 1023276.

Gupta, R., Kusuma, Y. S., & Prinja, S. (2020). Universal health coverage and older adults: A scoping review of current evidence. *Health Policy and Planning, 35*(10), 1290–1301.

Han, S. H., Tavares, J. L., Evans, M., Saczynski, J., & Burr, J. A. (2017). Social activities, incident cardiovascular disease, and mortality. *Journal of Aging and Health, 29*, 268–288.

Harvey, M., Rutherford, A., & Delany, A. (2020). Social contract theory and the rights of vulnerable populations: A literature review. *Health and Human Rights Journal, 22*(1), 211–226.

Hayotte, P., & Brunson, L. (2022). Becoming a grandparent in the internet age: An integrative model of intergenerational relationships within the family and grandparents' satisfaction with life. *Journal of Intergenerational Relationships*. Published online.

He, Q. H., Tan, Y. F., & Peng, Z. C. (2021). How does grandchild care affect the health of grandparents?—New evidence from CHARLS. *Population & Development (in Chinese), 27*(2), 52–64.

Hebert, L. E., Weuve, J., & Scherr, P. A. (2013). Alzheimer disease in the United States (2010–2050) estimated using the 2010 census. *Neurology, 80*(19), 1778–1783. https://doi.org/10.1212/WNL.0b013e31828726f5

Hofer, M., Hargittai, E., Büchi, M., & Seifert, A. (2019). Older adults' online information seeking and subjective well-being: The moderating role of internet skills. *International Journal of Communication, 13*, 4426–4443.

Hou, C., Lin, Y., Ren, M., Liu, M., Ma, Y., & Li, H. (2018). Cognitive functioning transitions, health expectancies, and inequalities among elderly people in China: A nationwide longitudinal study. *Geriatric Psychiatry, 33*(12), 1635–1644.

Hu, X., Gu, S., Zhen, X., Sun, X., Gu, Y., & Dong, H. (2021). Trends in cognitive function among Chinese elderly from 1998 to 2018: An age-period-cohort analysis. *Frontiers in Public Health*, 753871.

Huang, H., & Zhang, X. (2017). The adoption and use of WeChat among middle-aged residents in urban China. *Chinese Journal of Communication, 10*(2), 134–156.

Huxhold, O., Hees, E., & Webster, N. J. (2020). Towards bridging the grey digital divide: Changes in internet access and its predictors from 2002 to 2014 in Germany. *European Journal of Ageing, 17*, 271–280.

Inoue, Y., Wann, D. L., Lock, D., Sato, M., Moore, C., & Funk, D. C. (2020). Enhancing older Adults' sense of belonging and subjective well-being through sport game attendance, team identification, and emotional support. *Journal of Aging and Health, 32*(7–8), 530–542.

Jang, H., & Tang, F. Y. (2016). Effects of social support and volunteering on depression among grandparents raising grandchildren. *The International Journal of Aging, 83*(4), 491–507.

Japanese Cabinet. (1993). *Basic law for a longevity society.*

Jiang, X., & Zheng, Y. (2013). An analysis of the Main sources of financial support and social support system for Chinese elderly. *Journal of Population (Renkou Xuekan), 2*, 42–48.

Joe, J., & Demiris, G. (2013). Older adults and mobile phones for health: A review. *Journal of Biomedical Informatics, 46*, 947–954.
Kelsey, J. L., Berry, S. D., Procter-Gray, E., Quach, H. L., Uyen-Sa, D. T., Nguyen, D. S., Li, W. J., Kiel, D. P., Lipsitz, L. A., & Hannan, M. T. (2010). Indoor and outdoor falls in older adults are different: The maintenance of balance, independent living, intellect, and zest in the elderly of Boston study. *Journal of American Geriatric Society, 58*(11), 2135–2141.
Khalaila, R., & Vitman-Schorr, A. (2018). Internet use, social networks, loneliness, and quality of life among adults aged 50 and older: Mediating and moderating effects. *Quality of Life Research, 27*, 479–489.
Kim, J., Choi, Y., Choi, J. W., Nam, J. Y., & Park, E.-C. (2017). Impact of family characteristics by marital status of cohabitating adult children on depression among Korean older adults. *Geriatric Gerontology International, 17*(12), 2527–2536.
Lawrence, E., Mollborn, S., & Hummer, R. (2017). Health lifestyles across the transition to adulthood: Implications for health. *Social Science & Medcine, 193*, 23–32.
Lee, Y., Chang, Y., Chiang, T., Liu, C., & Shelley, M. (2020). Living arrangements and sleep-related outcomes among older adults in China: A panel analytic approach. *The International Journal of Aging and Human Development, 91*(2), 111–126.
Lee, Y., Kim, J., Back, J. H., Kim, S., & Ryu, M. (2013). Changes in combined lifestyle risks and disability transition in older adults: Korean longitudinal study of aging, 2006–2008. *Preventive Medicine, 56*, 124–129.
Li, L., Li, S., & Zhang, W. (2006). Intergenerational support and self-rated health of rural elderly. *Chinese Gerontology Studies (Zhongguo Laonianxue Zazhi), 26*(11), 1453–1455.
Li, L. W., Liu, J., Zhang, Z., & Xu, H. (2015). Late-life depression in rural China: Do village infrastructure and availability of community resources matter? *International Journal of Geriatric Psychiatry, 30*, 729–736.
Li, S., Fan, W., Zhu, B., Ma, C., Tan, X., & Gu, Y. (2022). Efects of age, period, and cohort on the prevalence of frailty in Chinese older adults from 2002 to 2014. *Frontiers in Public Health*, 935163.
Li, T., & Zhang, Y. (2023). Intergenerational relationships across multiple children and older parents' depressive symptoms in China—A resource contingency perspective. *Journals of Gerontology: Social Sciences*, Published online, *78*, 925–937.
Li, Y., Xu, L., Chi, I., & Guo, P. (2013). Participation in productive activities and health outcomes among older adults in urban China. *The Gerontologist, 54*(5), 784–796.
Liu, Q. (2014). Intergenerational relations and life satisfaction among Chinese male elderly. *Southern Population (Nanfang Renkou), 29*(4), 35–46.
Luo, M., Ding, D., Bauman, A., Negin, J., & Phongsavan, P. (2020). Social engagement pattern, health behaviors and subjective Well-being of older adults: An international perspective using WHO-SAGE survey data. *BMC Public Health, 20*, 99.
Lv, X., Li, W., Ma, Y., Chen, H., Zeng, Y., Yu, X., Hofman, A., & Wang, H. (2019). Cognitive decline and mortality among community-dwelling Chinese older people. *BMC Medicine, 17*, 63.
Martín-María, N., Caballero, F. F., Moreno-Agostino, D., Olaya, B., Haro, J. M., Ayuso-Mateos, J. L., & Miret, M. (2020). Relationship between subjective well-being and healthy lifestyle behaviours in older adults: A longitudinal study. *Aging & Mental Health, 24*(4), 611–619.
Martinez-Gomez, D., Guallar-Castillon, P., Garcia-Esquinas, E., Bandinelli, S., & Rodriguez-Artalejo, F. (2017). Physical activity and the effect of multimorbidity on all-cause mortality in older adults. *Mayo Clinic Proceedings, 92*, 376–382.
McDonnall, M. C. (2011). The effect of productive activities on depressive symptoms among older adults with dual sensory loss. *Research on Aging, 33*, 234–255.
Mollah, M. A. (2021). Aging policy analysis and evaluation. In D. Gu & D. M. Dupre (Eds.), *Encyclopedia of gerontology and population aging* (pp. 336–341). Springer Publisher.
Murtaugh, C., Spillman, B., & Wang, X. (2010). Lifetime risk and duration of chronic disease and disability. *Journal of Aging and Health, 23*(3), 554–577.
National Bureau of Statistics of China. (2022). *Statistical communiqué of the people's republic of china on the 2021 national economic and social development.*

National Center for Health Statistics. (2018). *Health, United States, 2018*. Retrieved from https://www.cdc.gov/nchs/data/hus/hus18.pdf

National Health Association of China. (2020). *China's hygiene and health yearbook*.

National Health Association of China. (2022). *China's hygiene and health yearbook*.

National Research Council, & Committee on Population. (2001). *Preparing for an aging world: The case for crossnational research*. National Academies Press.

Ng, S. T., Tey, N. P., & Asadullah, M. N. (2017). What matters for life satisfaction among the oldest-old? Evidence from China. *PLoS ONE, 12*(2), e0171799. https://doi.org/10.1371/journal.pone.0171799

OECD. (2019). *Pensions at a glance 2019: OECD and G20 indicators*. OECD Publishing.

Phulkerd, S., Gray, R., & Chamratrithirong, A. (2020). The influence of co-residential and non-co-residential living arrangements on sufficient fruit and vegetable consumption in the aging population in Thailand. *BMC Geriatrics, 20*, 476.

Rajan, K. B., Weuve, J., & Barnes, L. L. (2019). Prevalence and incidence of clinically diagnosed Alzheimer's disease dementia from 1994 to 2012 in a population study. *Alzheimers Dement, 15*(1), 1–7. https://doi.org/10.1016/j.jalz.2018.07.216

Ren, N., Li, Y., Wan, Z., & Wang, R. (2022). Patterns of HIV or AIDS mortality among older people from 1990 to 2019 in China: Age-period-cohort analysis. *JMIR Public Health and Surveillance, 8*(11), e35785.

Rizzuto, D., Orsini, N., Qiu, C., Wang, H. X., & Fratiglioni, L. (2019). Lifestyle, social factors, and survival after age 75: Population based study. *Journal of Aging and Health, 31*(7), 1297–1314.

Samanta, T., Chen, F., & Vanneman, R. (2015). Living arrangements and health of older adults in India. *The Journals of Gerontology Series B: Psychological Sciences and Social Sciences, 70*, 937–947.

Smith, J., Gerstorf, D., & Li, Q. (2008). Psychological resources for Well-being among octogenarians, nonagenarians, and centenarians: Differential effects of age and selective mortality. In Z. Yi, D. L. Poston, D. A. Vlosky, & D. Gu (Eds.), *Healthy longevity in China: Demographic, socioeconomic, and psychological dimensions* (pp. 329–346). Springer. https://doi.org/10.1007/978-1-4020-6752-5_20

Solé-Auró, A., & Crimmins, E. M. (2013). The oldest-old health in Europe and the United States. *Annual Review of Gerontology & Geriatrics, 33*(1), 1–33. https://doi.org/10.1891/0198-8794.33.3

Tsai, F.-J., Motamed, S., & Rougemont, A. (2013). The protective effect of taking care of grandchildren on elders' mental health? Associations between changing patterns of intergenerational exchanges and the reduction of elders' loneliness and depression between 1993 and 2007 in Taiwan. *BMC Public Health, 13*, 567.

United Nations. (2024). *World Population Prospects 2024*.

Wang, Z., Yang, X., & Chen, M. (2022). Inequality and associated factors in utilization of long-term care among Chinese older people: Evidence from the chinese longitudinal healthy longevity survey. *Social Indicators Research, 160*, 467–486. https://doi.org/10.1007/s11205-020-02279-x

Weaver, R. H., & Roberto, K. A. (2017). Home and community-based service use by vulnerable older adults. *The Gerontologist, 57*(3), 3540–3551.

Wei, K., Liu, Y., Yang, J., Gu, N., Cao, X., Zhao, X., Jiang, L., & Li, C. (2022). Living arrangement modifies the associations of loneliness with adverse health outcomes in older adults: Evidence from the CLHLS. *BMC Geriatrics, 22*, 59.

Weil, J., & Cabrera, C. (2021). Quality of life. In D. Gu & D. M. Dupre (Eds.), *Encyclopedia of gerontology and population aging* (pp. 4094–4101). Springer.

World Health Organization. (2017). *Depression and other common mental disorders: Global health estimates*.

World Health Organization. (2021). *Global health estimates 2020: Deaths by cause, age, sex, by country and by region, 2000–2019*.

Wu, Q., & Gu, D. (2021). Oldest-Old Adults. In D. Gu & D. M. Dupre (Eds.), *Encyclopedia of gerontology and population aging* (pp. 3637–3653). Springer.

Xinhua News Agency. (2016). *The communist party of China central committee and the state council promulgate the "Healthy China 2030 Planning Outline"*.

Xu, L., Wu, B., Chi, I., & Hsiao., H.-Y. (2012). Intensity of grandparent caregiving and life satisfaction among rural Chinese older adults: A longitudinal study using latent difference score analysis. *Family & Community Health, 35*, 287–299.

Yang, H., Zhang, S., Zhang, S. Q., Xie, L., Wu, Y., Yao, Y., Tang, L., & Li, Z. (2021a). Internet use and depressive symptoms among older adults in China. *Frontiers in Public Health, 12*, 739085.

Yang, S. Y., Xie, L. L., & Li, T. (2021b). *Living arrangements and intergenerational support among older Chinese: Does the gender of the co-resident adult children make a difference?* International Social Work, Published online.

Yang, Y. (2007). Is old age depressing? Growth trajectories and cohort variations in late-life depression. *Journal of Health and Social Behavior, 48*, 16–32.

Yi, Z., & Vaupel, J. W. (2002). Functional capacity and self–evaluation of health and life of oldest-old in China. *Journal of Social Issues, 58*(4), 733–748. https://doi.org/10.1111/1540-4560.00287

Yu, R., Wong, W., Chong, K., Chang, B., Lum, C., Auyeung, T., Lee, J., Lee, R., & Woo, J. (2018). Trajectories of frailty among Chinese older people in Hong Kong between 2001 and 2012: An age-period-cohort analysis. *Age & Aging, 47*(2), 254–261.

Yuan, H. (2021). Internet use and mental health problems among older people in Shanghai, China: The moderating roles of chronic diseases and household income. *Aging & Mental Health, 25*(4), 657–663.

Zeng, Y., Feng, Q., & Hesketh, T. (2017). Survival, disabilities in activities of daily living, and physical and cognitive functioning among the oldest-old in China: A cohort study. *Lancet, 389*(10079), 1619–1629. https://doi.org/10.1016/S0140-6736(17)30548-2

Zeng, Y., Land, K. C., & Gu, D. (2014). Effects of changes in household structure and living arrangements on future home-based care costs for disabled elders in the United States. In Y. Zeng, K. C. Land, & D. Gu (Eds.), *Household and living arrangement projections: The extended cohort-component method and applications to the U.S. and China* (pp. 167–188). Springer.

Zhang, H., Wang, H., Yan, H., & Wang, X. (2022). Impact of internet use on mental health among elderly individuals: A difference-in-differences study based on 2016–2018 CFPS data. *International Journal of Environmental Research and Public Health, 19*, 101.

Zhang, J. (2016). Aging in cyberspace: Internet use and quality of life of older Chinese migrants. *The Journal of Chinese Sociology, 3*(1), 26.

Zhang, J., Han, A., & Wu, X. (2021). Relation of depressive symptoms to intergenerational support and internet use in empty nest elderly. *Chinese Mental Health Journal, 35*(10), 838–843.

Zhang, L. (2021). A latent class analysis of health lifestyles and health outcomes among Chinese older adults. *Ageing & Society, 41*, 313–338.

Zhang, N., Bécares, L., Chandola, T., & Callery, P. (2015). Intergenerational differences in beliefs about healthy eating among carers of left-behind children in rural China: A qualitative study. *Appetite, 95*, 484–491.

Zhang, Q., & Li, Z. (2022). The impact of internet use on the social networks of the elderly in China—The mediating effect of social participation. *International Journal of Environmental Research and Public Health, 19*, 9576.

Zhang, S., & Zhang, Y. (2021). The relationship between internet use and mental health among older adults in China: The mediating role of physical exercise. *Risk Management and Healthcare Policy, 14*, 4697–4708.

Zhang, W., & Chen, M. (2014). Psychological distress of older Chinese: Exploring the roles of activities, social support, and subjective social status. *Journal of Cross-Cultural Gerontology, 29*(1), 37–51.

Zhang, X., & Silverstein, M. (2022). Intergenerational emotional cohesion and psychological Well-being of older adults in rural China: A moderated mediation model of loneliness and friendship ties. *The Journals of Gerontology. Series B, Psychological Sciences and Social Sciences, 77*(3), 525–535.

Zhang, Z. (2004). The impact of adult children providing care on the elderly's health: Promotion or selection (Zinu Zhaoliao Dui Laonianren Jiankang De Yingxiang: Cuijin Haishi Xuanze)? *Chinese Population Sciences (Zhongguo Renkou Kexue), Supplement*, 29–36 (in Chinese).

Zhao, Y., Dupre, M. E., Qiu, L., & Gu, D. (2017). Changes in perceived uselessness and risks for mortality: Evidence from a national sample of older adults in China. *BMC Public Health, 17*, 561.

Zhou, J., Mao, W., Lee, Y., & Chi, I. (2017). The impact of caring for grandchildren on grandparents' physical health outcomes: The role of intergenerational support. *Research on Aging, 39*(5), 612–634.

Zhou, J., & Walker, A. (2016). The need for community care among older people in China. *Aging & Society, 36*, 1312–1332.

Review of Literature

2

Chapter 1 has briefly pointed out that health related issues have received considerable research attention in multiple disciplines. This chapter will provide an in-depth review of literature on older adults' and the oldest old's health, with a particular focus on sociological and demographic studies of health of older adults, particularly the oldest old. The following sections of this chapter will review several mainstreams of existing studies. Merits and limitations of previous studies will be discussed. The chapter also discusses how this manuscript advances our understanding of Chinese oldest old's health and fills the gaps of previous literature.

2.1 Studies of Older Adults' Health from Multiple Disciplines

A group of existing health studies have focused on evaluating health status of older adults by paying attention to physical and physiological indicators of older adults. Research on physiological health has investigated aspects such as metabolism, bone health, vision and hearing of seniors. For example, as early as 1985, Tylavsky and associates conducted a study on the bone health of older adults. The results emphasized the importance of bone health for overall physiological well-being of older adults and pointed out the importance of calcium intake in maintaining their health (Tylavsky et al., 1985). In following years, researchers have extended this line of research and added new findings on calcium intake, bone mineral density and older adults' health (Adatorwovor et al., 2015; John & Philip, 2013; Kim et al., 2019).

Those who studied functional status of older adults have examined cognitive function, immune function, and cardiorespiratory function of older adults. Scholars have observed and used indicators such as memory, attention and processing speed when assessing cognitive abilities of elders. Jerger and colleagues, for example, studied the relationship between speech audiometry scores and cognitive status among 200 seniors by analyzing degrees of pure-tone sensitivity loss.

L. Zhang, *Living Longer and Healthier at Older Ages*, Advances in Studies of Aging and Health 3, https://doi.org/10.1007/978-3-031-69773-9_2

They demonstrated that speech audiometry scores could effectively predict cognitive status of older adults (Jerger et al., 1991). Later on, researchers enriched this group of studies by emphasizing the importance of physical exercise in promoting older adults' cognitive function (Dostálová et al., 2021; Wei et al., 2021a, b).

Other scholars studied immune function of older adults by evaluating immune cells and related inflammation levels among older adults. They pointed out that the deterioration immune function due to aging has a negative impact on older adults' overall physiological health. Also, immunological decline has been found to be associated with depression and stress, which damages older adults' health outcome (Applegate et al., 2002). This group of work also touched on issues such as immune health among HIV-positive older adults (Rendina et al., 2017), influenza vaccine and older adults' health (Shapiro et al., 2020) and immune response to COVID-19 among older adults (Chen et al., 2021; Jergovic et al., 2021).

Research on cardiorespiratory function has observed indicators such as lung capacity, lung elasticity, heart rate and related cardiorespiratory fitness to assess physiological health of older adults (Brown et al., 2019; Li et al., 2017a). Physiological research on chronic diseases of older adults has primarily focused on studying the prevalence, risk factors and management of chronic diseases, including heart disease, hypertension, diabetes et al. (Alves et al., 2021; Bakouni et al., 2017; Dolores et al., 2017; Lu et al., 2021).

Psychological research of older adults' mental health has also been a mainstream of existing health studies. This group of analyses have examined issues such as anxiety, feelings of loneliness, low mood, depression, suicidal tendencies et al. These studies can be categorized into several groups. The first group focused on exploring the interactions of various psychological health-related indicators. For instance, researchers investigated the relationship between loneliness, depression, anxiety, emotions and suicide (Fernandez-Rodrigues et al., 2022; Kim et al., 2009; Lin et al., 2014a; Thalmann et al., 2010). The second group of research explored social factors that may possibly influence older adults' psychological health. These factors included social support, family cohesion, lifestyles et al. (Reiland et al., 2021; Zhou et al., 2020). The third group of studies concentrated on investigating treatments of older adults' mental illnesses at individual, family and societal levels (Grando & Grando, 2020; Perng & Renz, 2017).

2.2 Demographic and Social Determinants of Older Adults' Health

After reviewing physiological and psychological studies of older adults' health, this section moves to reviewing sociological and demographic studies of older adults' health and its determinants. Demographic, socioeconomic and a variety of social factors, including social support, health lifestyles, internet use, grandparental childcare, living arrangements, social participation and environmental factors, have been proved to have significant impacts on older adults' health. This section reviews findings on how these factors have impacted older adults' health.

Demographic factors such as sex, age, race and ethnicity, marital status, urban-rural residence, immigration/migration status have been repeatedly shown to be major determinants of older adults' health (Du & Xu, 2016; Ellison, 1995; Fülöp et al., 2017; Glei et al., 2013; Kawachi et al., 2005; Laaksonen et al., 2002; Zhu et al., 2022). Socioeconomic status (SES), usually measured by income, educational attainments, occupation, has also been found to have significant influence on individual health. Individuals who have higher SES tend to live longer and report better health as compared to those with lower SES (Chen et al., 2010; Kim et al., 2004; Lynch, 2003; Pampel et al., 2010; Zimmer et al., 2004). Previous studies also highlighted the cumulative effects of SES on individual health, especially in older ages (Luo & Waite, 2005; Zhang, 2021).

Besides demographic and socioeconomic factors, social support and social networks including online networks have been found to be important covariates that promote older adults' health (Bai et al., 2020; Cohen & Janicki-Deverts, 2009; Jang & Tang, 2016; Nabi et al., 2013; Petrovcic et al., 2015). As a main form of social support, intergenerational support has been documented to play an extremely vital role in shaping older adults' health status. Intergenerational support contains emotional, instrumental and economic support. Research indicated that informal support provided by adult children can significantly improve older adults' mental health condition, subjective wellbeing and life satisfaction (Giles et al., 2011; Inoue et al., 2020; Li et al., 2006; Zunzunegui et al., 2001). Seniors with family functional impairments have been revealed to have worse physical and mental health as compared to those who had intact families (Chen & Silverstein, 2000; Kim & Kim, 2003; Levitt et al., 1992; Li et al., 2006). Scholars further pointed out that stronger intergenerational ties built through internet use may benefit health of their elder parents (Lv & Zhang, 2021). In terms of economic support, findings showed that the impact of economic support from adult children on older adults' health varied depending on older adults' income levels. Economic support from adult children did not have a significant effect on health of older adults with higher income; a significant effect was only shown among those with low and moderate income levels (Chen et al., 2020).

In addition to intergenerational support, government and community support has also showed significant influence on seniors' health. In China, government support has long been considered as a key factor for regional development and health promotion among community-dwelling citizens (Chen & Wu, 2016). A transition has been observed in terms of older adults using home- and community-based services (Chen & Berkowitz, 2012; Leung, 2010; Tian & Wang, 2014). The Chinese government is now highly encouraging home-based community care for elders. Researchers found a strong correlation between older adults' health status and the level of community support (Yu et al., 2018b). They emphasized that formal services proved by government has played a critical role in fulfilling needs of community-dwelling older adults and improving their general health condition (Chen & Wu, 2016; Lafortune et al., 2009; Tian & Wang, 2014).

The third domain of influential factors documented in prior literature is healthy lifestyles. Healthy lifestyles help seniors to maintain a healthy physiological and psychological state. The main aspects of healthy lifestyles consist of healthy dietary patterns, moderate exercise, no smoking, no drinking and good sleep (Burdette et al., 2017; Chai & Mei, 2022; Chan & Leung, 2015). Elders who experienced sleep disorders were shown to have a higher risk of being sick than others with good sleep quality (Fadayevatan et al., 2020; Liu et al., 2019). Prior research also indicated that as dietary quality improves, physical health of older adults also improves, suggesting a positive correlation between dietary quality and health of older adults (Gezer et al., 2020). Others considered dietary quality as a moderator in the relationship between sleep quality and physical and mental health (Zhao et al., 2021). Studies suggested that elder women with good dietary quality tended to have stronger bones as compared to those with poor dietary quality, suggesting a positive association between dietary quality and bone health among elder women (Destefani et al., 2021). Depression among older adults has also been found to be linked to people's dietary quality. Individuals with poor dietary patterns and styles were more likely to suffer from depression as compared to those with good dietary quality (De Oliveira Meller et al., 2021). Unhealthy habits such as smoking and drinking have been found to have significantly negative impacts on older adults' cognitive function with smokers and drinkers showing a slower brain processing speed than non-smokers and non-drinkers (Fadayevatan et al., 2020).

With age increasing, older adults' physical activities decline. Physical exercise has been documented to be able to maintain health status of older adults (Lampek et al., 2021). Recent studies investigated physical exercise and older adults' health during the COVID-19 pandemic and showed that reduction in regular exercise was significantly correlated with a worse health condition and lower quality of life (Tamai et al., 2022). Researchers have also explored how specific types of exercise linked to health status of older adults. Chao and associates argued that tennis exercise had a positive effect on cardiovascular metabolic health parameters among seniors (Chao et al., 2021).

Prior literature further emphasized the importance of living arrangements on influencing older adults' health and well-being. Some researchers showed a positive associated between co-residence and older adults health (Hughes & Waite, 2002; Lawton et al., 1984; Magaziner et al., 1988; Sereny & Gu, 2011; Zimmer, 2005). They argued that the household provides an important context to individuals for social integration and various dimensions of support. Co-residence with adult children meets the needs of daily care and provides physical and emotional support to older adults. Prior literature showed that elders who lived with others had the lowest mortality rate (Lund et al., 2002). Those who lived in institutions were found to have a higher mortality rate (Feng et al., 2017). Older adults who lived alone were also found to have poor health, including depression, cardiovascular disease and dementia (Desai et al., 2020; Gan et al., 2021). Other studies, however, challenged the above findings by contending that co-residence damages older adults health and subjective well-being (Rook, 1984; Rook & Pietromonaco, 1987). They argued that the negative impact of co-residence on older adults' health and subjective well-being may be due to the fact that relations among household members can create tensions, conflicts, and negative interaction patterns, which in turn harms their well-being (Krause & Liang, 1993; Rook, 1984). Empirical evidence showed living alone benefited older adults' mental health (Kim et al., 2017). Older adults who lived alone tended to be healthier than those who lived with others (Michael et al., 2001). Thus, the relationship between living arrangements and older adults' health and subjective well-being is not conclusive.

In China where there is a strong tradition of filial piety and an expectation of sons providing care to elder parents, empirical findings largely corroborated a positive effect of co-residence on seniors' health. In general, older adults who lived with adult children were found to report better SRH, a lower likelihood of having activity of daily living (ADL) disability and were less likely to feel lonely or depressed (Chen et al., 2015; Wei et al., 2022; Xu et al., 2022; Zhou et al., 2018). Chinese older adults who lived with family members also tended to report better quality of sleep and a longer duration of sleep (Lee et al., 2019).

Grandparental child care is the fifth important factor that has been repetitively shown to have significant effects on older adults' health. Previous studies have found negative, positive and even "U" shape associations between grandchild care and grandparental health. Those who supported a negative relationship argued that grandparenting is time consuming and is physically and psychologically demanding for grandparents. It is generally associated with a decrease in time available for hanging out with friends and doing recreational activities. Some grandparents may have given up employment to raise grandchildren. Custodial grandmothers have been found to be more likely economically disadvantaged because job and income loss increased their economic vulnerability and psychological stress (Jang & Tang, 2016). In addition, lacking of privacy, social engagement and leisure time often led to heightened isolation and depression (Baker & Merril, 2008), which brought consequently negative effects on grandparents' health (Ates, 2017).

Others who supported a positive association between grandparental child care and older adults' health contended that caring for a grandchild may have resulted in more active lifestyles, healthier meals and a reduction in smoking (Peng & Chan, 2020; Waldrop & Weber, 2001). It has made grandparents stay active and has prevented elders from deteriorating at older ages (Arpino & Bordone, 2014). Taking care of grandchildren could also stimulate the brain, which is helpful in maintaining good cognitive skills (Engelhardt et al., 2010). Meanwhile, grandparenting can enhance grandparents' self-esteem, self-worth as well as family cohesion (Di Gessa et al., 2016; Hilbrand et al., 2017b). Grandchildren could be great emotional resources for grandparents, which led to reduced depressive symptoms (Tsai et al., 2013) and a greater level of life satisfaction (Ku et al., 2013; Xu et al., 2012). Moreover, caregiving to grandchildren was found to be able to reduce grandparents' mortality rates (Hilbrand et al., 2017a).

Some prior studies further showed no significant association (Hughes et al., 2007; Oshio, 2022) or a nonlinear relationship between grandparenting and grandparental health. Glaser et al. (2014) claimed that non-intensive levels of caregiving were most likely to be associated with better health. Grandparents who provided custodial childcare and grandparents who did not provide any grandchild care were likely to report poor health (Glaser et al., 2014). The inconsistent findings on caring for grandchildren and grandparental health shown in the existing literature may be explained by methodological issues, different social-cultural background and the studying subjects.

The next social factor that has been found to have a strong link to older adults' health is internet use. With the prevalence of internet use worldwide, the effects of internet use on older adults' health has become a hot topic in health studies. Internet use was found to be able to enlarge older adults' social networks and increase the level of social support, including intergenerational support. Studies examining sample from Western countries showed that using internet increased social contact between seniors and other people, which helped older adults receiving more support from adult children and others. Consequently, it promoted elders' health status (Jolie & Lam, 2006; Swickert & Hittner, 2002). It was documented that internet use extended social networks of seniors by improving the level of their social participation (Zhang & Li, 2022). Scholars also provided evidence that using internet increased intergenerational contact, which helped older adults receiving more intergenerational support from their adult children (Li, 2012; Lv & Zhang, 2021). Researchers also showed that internet use can reduce depressive symptoms among older adults by increasing their social engagement levels, creating contact with social ties and connecting with others (Chen & Schulz, 2016; Heo et al., 2015; Yu et al., 2016a). As a result, social engagement through internet use benefited seniors' cognitive function and subjective well-being (Plaza et al., 2011). Further, online networks built by internet use were able to provide seniors various forms of support and entertainment which may not be obtained from daily life (Tsai et al., 2015; Veena et al., 2012). Internet and smart phone usage helped older adults in terms of health maintenance, chronic disease self-management, communication with professionals, monitoring health status, and receiving health promotion interventions (Joe & Demiris, 2013; Lee & Carlisle, 2011). Using internet and smart phones have been found to be able to support daily life activities by functioning as navigators and memory aids, which preserved seniors' independence and improved quality of life (Plaza et al., 2011).

Excessive or compulsory use of internet, nevertheless, decreased the time older adults spent with friends and lowered the level of in-person local social networking, which increased loneliness and lowered quality of life (Muusses et al., 2014). Problematic social media use was found to be linked to reduced social and psychological well-being (Marttila et al., 2021). Thus, previous studies showed both sides of internet use when it comes to its association with older adults' health.

Environmental issues have also been a concern of previous studies on older adults' health. In the Internet era, the online environment obviously affects health status of seniors. Research has shown that when interacting, communicating, and obtaining information from the online environment, positive emotions are stimulated to help seniors to maintain a comfortable emotional state, which benefits seniors' mental health (Hu et al., 2017). Physical environment is crucial as well. Prior research suggested a negative correlation between temperature and blood pressure (Song et al., 2022). The incidence of respiratory diseases among older adults was also believed to be closely related to climate characteristics. Humid and foggy climates can exacerbate respiratory diseases among seniors (Gazerro et al., 1996). Researchers also highlighted that older adults' visual impairments can be affected by their living environment (Siira et al., 2022).

In sum, previous analyses have showed strong links between older adults' health and a variety of demographic, socioeconomic, genetic, environmental and social factors. This chapter is not able to exhaust all of them. But obviously the majority of the existing studies are based on analyzing cross-sectional data. Only a limited number of studies have applied a longitudinal scope to examine older adults' health and its determinants. In the next section of the chapter, I briefly review studies of older adults' health from a longitudinal scope.

2.3 Studies of Older Adults' Health from a Longitudinal Scope

As compared to cross-sectional analyses, studies that applied a longitudinal scope, including the age-period-cohort (APC) approach, to examine older adults' health and its determinants are relatively inadequate. Among these studies, the majority of them focused on studying older adults' health in the U.S., Canada, Japan and other more developed countries. These studies can be classified as follows: First, investigating disability trends among the U.S. older adults. Results indicated that the period trends showed a substantial decline in instrumental activity of daily living (IADL) disability, and activity of daily living (ADL) disability remained stable over time. Unadjusted cohort trends for IADL and ADL disability also showed continual declines across successive cohorts. However, increasing cohort trends were evident in the adjusted models (Lin et al., 2012a, b, 2014b).

Second, examining age, period and cohort effects in depression prevalence among older adults (Abrams & Mehta, 2018; Sullivan et al., 2019; Xiang & Cheng, 2019; Yang & D'Arcy, 2023; Zivin et al., 2013). Results showed growth curves of depression over age among older adults. Younger cohorts showed an increasing trend in having depression symptoms. Overall late-life depressive symptom burden declined significantly over time. This decrease appeared to be driven primarily by greater reductions in depressive symptoms among the oldest old, and by an increase in those with no depressive symptoms.

Third, studying age, period and cohort effects in other health related trajectories. For instance, scholars examined trends in grip strength and investigated the age, period, and cohort effects on grip strength among older adults from Germany, Sweden, and Spain (Beller et al., 2019; Strand et al., 2019). Others examined older adults' perceived loneliness by time period and across birth cohorts. They showed that overall, loneliness decreased with age through the early 70s, after which it increased gradually. Researchers contended that loneliness is associated with poor health, living alone or without a spouse-partner, and having fewer close family and friends. All these factors together accounted for an overall increase in loneliness after age 75 (Hawkley et al., 2019).

As to studies of Chinese older adults, scholars have investigated the following subjects and documented the age, period and cohort effects. First, prior literature used the APC approach to scrutinize mortality trends caused by diabetes, liver, esophageal and gastric cancer (Guo & Li, 2012; Li et al., 2017b; Liu et al., 2020; Sun et al., 2018). Age, period and cohort effects were documented in prior literature. Taking the piece studying lung cancer mortality as an example, the results indicated that the lung cancer mortality attributable to smoking was generally increasing among Chinese adults over time. The results suggested that rapid aging and prevalence of smoking may intensify the increasing mortality of lung cancer in China.

The second group of literature investigated the effects of age, period, and cohort on frailty prevalence among older adults in Hong Kong and mainland China (Li et al., 2022; Yu et al., 2018a). This group of studies also examined trends in Chinese older adults' cognitive function (Hu et al., 2021) and disability (Hu et al., 2022; Liang et al., 2017; Pan et al., 2023; Yu et al., 2016b). The findings pointed out that ADL disability increased with age. A V-shaped cohort trend and a fluctuated period trend were identified by studying the Chinese sample. Females and urban residents were associated with more ADL limitations. When age increased, the gender and residence gaps in disability further increased. The cohort-based gender and residence inequalities in ADL limitations converged with successive birth cohorts. The age effect on trends in ADL disability was stronger as compared to period and cohort effects. Moreover, the gender and residence disparities in disability increased with age and decreased with successive cohorts.

The third aspect of prior longitudinal studies dealt with trajectories of older adults' health and mortality by highlighting the age, period and cohort effects. For example, Wei and colleagues studied trajectories of suicide rates among Chinese older adults aged 60–84 during 2003–2018 (Wei et al., 2023). They claimed that the suicide rates increased with age. The period effect was more pronounced among urban seniors, which showed an "U" shape with time change. No significant gender differences were observed. The cohort effect was significant since earlier rural birth cohorts showed higher suicide rates than their urban counterparts and the reversed pattern seemed to be true for later birth cohorts. Through analyzing the Chinese Cancer Registry Annual Report published in 2008–2019, Guo and associates studied cancer incidence among Chinese older adults aged 60 and above (Guo et al., 2022). They showed that cancer incidence increased with age regardless of sex and urban and rural residence. The period effect increased from 2005 to 2010. It then decreased from 2010 to 2015 and the downward trend was more obvious since 2005. The incidence rate among later-born cohorts was lower. Others also applied the APC method to examine patterns of HIV or AIDS mortality among older Chinese from 1990 to 2019 in China (Ren et al., 2022). These analyses have largely improved our understanding of the age, period and cohort effects in trajectories of various health measures among Chinese older adults.

2.4 Research on the Oldest Old's Health

Academic studies on the oldest old population started from the 1970s and nourished in the 1980s (Lund & Wang, 2020). The book *The Oldest-Old* (Suzman et al., 1992) has been considered as an important milestone in the field because after this publication, studies on the oldest old increased by over 100% in 3 years (Lund & Wang, 2020). According to Wu and Gu (2021), from 1991 to April 1, 2019, there were 4437 academic articles about the oldest old population in the Web of Science (WoS) database. The articles came from the fields of gerontology, epidemiology, physiology, sociology, and psychology. The top 25 authors (with most publications) were mainly from the USA, China, Italy and Spain. The top journals for the oldest old research include *Journal of the American Geriatrics Society, The Journal of Gerontology, Archives of Gerontology and Geriatrics, and Experimental Gerontology.*

By looking at the existing literature on the oldest old, Wu and Gu (2021) also pointed out that although the major keywords did not substantially change during the past few decades, there was a degradation of research focusing on gender differences, ADLs, dementia, and the United States. But there was an upgradation of research on longevity, China, and linkages with young adults. In addition, studies applying the life course approach emerged. Genetic studies and studies on treatment outcomes have been growing as well.

Considering a high level of morbidity and mortality among the oldest old, one main stream of prior studies has been focusing on the oldest old's health and its determinants. Regarding health determinants, previous research found a strong link between individual level socioeconomic status (SES) and the oldest old's disability, cognitive impairment and mortality (Zeng et al., 2017; Zhu & Xie, 2007). These findings supported theories of persistent disparity at older ages, i.e. the divergence, convergence, and persistent inequality theories in health caused by SES differentials. Using a group-based modeling approach to jointly estimate disability and mortality trajectories over time based on analyzing Chinese elders aged 80 and above, Zimmer and associates (2012) further identified three hierarchical groups of disability trajectory for each sex: i.e., starting low and rising gradually, starting low and rising quickly, and starting moderate and remaining stable. They pointed out that the trajectory was linked to individual SES characteristics and varied by sex.

Scholars also indicated the role of health lifestyles in influencing the oldest old's health. It was found that among Chinese oldest old, daily intake of vegetables and fruits (FV) was associated with 15.0–25.0% lower mortality risk as compared to seldom or never intake FV (Shi et al., 2015). Active lifestyle was found to have a positive effect on the oldest old's health as well. It was shown that frequent involvement in leisure activities was associated with 10.0–20.0% lower mortality among Chinese oldest old (Li et al., 2020). Although participating in religious activities, a way of social participating, was documented to have a stronger effect on health of young old men than on health of the oldest old men (Zeng et al., 2011), the oldest old who were involved in religious activities were found to have a lower mortality rate (Zhang, 2008).

A number of other factors have also been found to be influential on oldest old's health. For instance, previous research examined the associations between psychological resilience or Positive Affect and Negative Affect scale (PANAS) and the oldest old's health. Relying on a fuzzy set analysis, Gu and Feng (2018) indicated that resilience was linked to lower mortality, normal cognitive function, free of disability and comorbidity, better self-rated health (SRH) and life satisfaction. Hao et al. (2020) also found that adequate access to health care could increase the life expectancy by 2.0–2.5 years for older adults aged 65 and by 1.0–1.2 years for the oldest old. Genetic and gene-environment research further showed that genes can affect longevity. Through analyzing genotypic/phenotypic data from 1817 centenarians and younger individuals, scholars demonstrated that three SNPs on the FOXO1A gene and three SNPs on the FOXO3A gene were related to longevity traits in Han Chinese populations (Zeng et al., 2010). The gene-environmental interaction between FOXO genotypes and tea drinking was also found to be significantly associated with health of older adults. These findings implied that the health benefits of nutritional interventions, such as tea drinking, may rely upon and interact with individual genetic profiles (Zeng et al., 2015). Additionally, the genetic associations with longevity was documented to vary by sex. Through analyzing CLHLS genotypic/phenotypic data, prior research found that 11 male-specific and 11 female-specific longevity loci and 35 male-specific and 25 female-specific longevity loci were correlated with longevity among Chinese seniors (Zeng et al., 2018).

The existing studies on the oldest old have enriched our understanding about the mechanisms of aging and healthy longevity. Nonetheless, some areas are still overlooked or understudied, which deserve more research attention in future. First, cross-national comparative studies are limited due to unavailability of survey data on the oldest old in some countries, particularly developing countries. Even for countries with available data on the oldest old, data accuracy is still an issue, especially for less developed nations. It prevents researchers from conducting comparative studies of the oldest old's health, morbidity and mortality due to misreporting of age and health condition. Second, studies of compound effects of period and cohort on the oldest old's health are limited. Older adults, especially the oldest old are more vulnerable to climate change, natural disasters, social events, and epidemics/pandemics. Previous analyses have not adequately addressed the impacts of social events, i.e. period effects, on the oldest old's health and well-being. Meanwhile, the birth cohort effects have not been paid enough attention either. Examining the compound effects of period and cohort on individual health in oldest-old ages should also be a focus of future research. Lastly, as Wen and Gu (2021) pointed out, even with a growing body of research using a holistic model to study longevity and its determinants, such applications are still rare in the field. Building biodemographic models are warranted to better understand aging mechanisms. Such research will also have important policy implications for addressing population aging and related challenges.

2.5 Summary and an Evaluation of Prior Literature

To summarize, this chapter has reviewed previous studies on health and its influential factors among older adults, including the oldest old. Attention is given to demographic and sociological studies. Previous work has significantly advanced our understanding of older adults' health, particularly health of the oldest old. Empirical findings also contained important policy implications in terms of promoting seniors' health and subjective wellbeing. Nonetheless, most existing studies applied a cross-sectional scope, which prevents us from understanding the changing patterns of seniors' health and its influential factors. Although some studies have applied a longitudinal view to carry out analyses, these studies only touched on limited health related subjects. The compound effects of age, period and cohort are still unclear when linking some important demographic, behavioral and social factors to the oldest old's health, which remains a gap in the existing literature. This manuscript thus aims to improve previous literature by filling such voids. Data and methods that are used by this research to conduct empirical work will be discussed in next chapter.

References

Abrams, L., & Mehta, N. (2018). Depression trajectories in older adults: Differences by gender, race/ethnicity, education and cohort. *Innovation in Aging, 23*, 1797.

Adatorwovor, R., Roggenkamp, K., & Anderson, J. J. (2015). Intakes of calcium and phosphorus and calculated calcium-to-phosphorus ratios of older adults: NHANES 2005–2006 data. *Nutrients, 7*(11), 9633–9639.

Alves, V., de Oliveira, R., & Gregori, D. (2021). Non-communicable chronic diseases: Mortality of older adult citizens in Brazil and Italy before the Covid-19 pandemic. *Health Policy OPEN, 2*, 100041.

Applegate, K., Kiecolt-Glaser, J., & Glaser, R. (2002). Depression, immune function, and health in older adults. *Physical Illness & Depression in Older Adults, 68177148*. https://doi.org/10.1007/0-306-47178-7_7

Arpino, B., & Bordone, V. (2014). Does grandparenting pay off? The effect of child care on grandparents' cognitive functioning. *Journal of Marriage and Family, 76*, 337–351.

Ates, M. (2017). Does grandchild care influence grandparents' self-rated health? Evidence from a fixed effects approach. *Social Science & Medcine, 190*, 67–74.

Bai, Y., Bian, F., Zhang, L., & Cao, Y. (2020). The impact of social support on the health of the rural elderly in China. *International Journal of Environmental Research and Public Health, 17*(6), 2004.

Baker, L. A., & Merril, S. (2008). Preventive health behaviors among grandmothers raising grandchildren. *Journals of Gerontology, 63*(5), 304–311.

Bakouni, H., Guerra, S., Chudzinski, V., Berbiche, D., & Vasiliadis, H. (2017). One-year prospective study on the presence of chronic diseases and subsequent cognitive decline in older adults. *Journal of Public Health, 294*, 357–365.

Beller, J., Miething, A., Regidor, E., & Lostao, L. (2019). Trends in grip strength: Age, period, and cohort effects on grip strength in older adults from Germany, Sweden, and Spain. *SSM – Population Health, 21*(9), 100456.

Brown, B., Castalanelli, N., & Rainey-Smith, S. (2019). Influence of BDNF Val66Met on the relationship between cardiorespiratory fitness and memory in cognitively normal older adults. *Behavioural Brain Research: An International Journal, 2*, 1123–1129.

Burdette, A., Needham, B., Taylor, M., & Hill, T. (2017). Health lifestyles in adolescence and self-rated health into adulthood. *Journal of Health and Social Behavior, 58*, 520–536.
Chai, X., & Mei, J. (2022). Investigating food insecurity, health lifestyles, and self-rated health of older Canadians living alone. *BMC Public Health, 22*, 2264.
Chan, C. W., & Leung, S. F. (2015). Lifestyle health behaviors of Hong Kong Chinese: Results of a cluster analysis. *Asia-Pacific Journal of Public Health, 27*(3), 293–302.
Chao, H., Liao, Y., & Chou, C. (2021). Influences of recreational tennis-playing exercise time on cardiometabolic health parameters in healthy elderly: The ExAMIN age study. *International Journal of Environmental Research and Public Health, 18*(3), 1255.
Chen, Y. M., & Berkowitz, B. (2012). Older adults' home- and community-based care service use and residential transitions: A longitudinal study. *BMC Geriatrics, 12*, 44.
Chen, Y. R. R., & Schulz, P. J. (2016). The effect of information communication technology interventions on reducing social isolation in the elderly: A systematic review. *Journal of Medical Internet Research, 18*, 1–11.
Chen, X., & Silverstein, M. (2000). Intergenerational social support and the psychological well-being of older parents in China. *Research on Aging, 22*(1), 43–65.
Chen, Y., & Wu, L. (2016). Characteristics of public health service resources and government responsibility for mental health of rural elderly. *Qiusuo, 12*, 77–81.
Chen, F., Yang, Y., & Liu, G. (2010). Social change and socioeconomic disparities in health over the life course in China: A cohort analysis. *American Sociological Review, 75*(1), 126–150.
Chen, W., Fang, Y., Mao, F., Hao, S., Chen, J., Yuan, M., & Hong, Y. A. (2015). Assessment of disability among the elderly in Xiamen of China: A representative sample survey of 14,292 older adults. *PLoS One, 10*, e0131014.
Chen, G., Si, W., & Qiu, L. (2020). Intergenerational financial transfers and physical health of old people in rural China: Evidence from CHARLS data. *Ciência Rural, 50*, 5.
Chen, Y., Klein, S., Garibaldi, B., Li, H., & Wu, C. (2021). Aging in COVID-19: Vulnerability, immunity and intervention. *Aging Research Review, 65*, 101205.
Cohen, S., & Janicki-Deverts, D. (2009). Can we improve our physical health by altering our social networks? *Perspectives on Psychological Science, 4*, 375–378.
De Oliveira Meller, F., Manosso, L., & Schäfer, A. (2021). The influence of diet quality on depression among adults and elderly: A population-based study. *Journal of Affective Disorders, 282*, 1076–1081.
Desai, R., John, A., Stott, J., & Charlesworth, G. (2020). Living alone and risk of dementia: A systematic review and meta-analysis. *Ageing Research Review, 62*, 101122.
Destefani, S., Kurokawa, C., Rodrigues, S., Corrente, J., Padovani, C., Paiva, S., & Mazeto, G. (2021). Is there a relationship between diet quality and bone health in elderly women? A cross-sectional study. *Archives of Endocrinology and Metabolism, 65*, 609–616.
Di Gessa, G., Glaser, K., & Tinker, A. (2016). The impact of caring for grandchildren on the health of grandparents in Europe: A lifecourse approach. *Social Science & Medicine, 152*, 166–175.
Dolores, M., Hortensia, R., & Svetlana, D. (2017). Multimorbidity patterns in older adults: An approach to the complex interrelationships among chronic diseases. *Archieves of Medical Research, 1*, 160–167.
Dostálová, R., Stillman, C., Erickson, K., Slepika, P., & Mudrák, J. (2021). The relationship between physical activity, self-perceived health, and cognitive function in older adults. *Brain Sciences, 11*(4), 492.
Du, Y., & Xu, Q. (2016). Health disparities and elayed health care among older adults in California: A perspective from race, ethnicity, and immigration. *Public Health Nursing, 33*(5), 383–394.
Ellison, C. G. (1995). Race, religious involvement and depressive symptomatology in a Southeastern U.S. Community. *Social Science & Medicine, 40*(11), 1561–1572.
Engelhardt, H., Buber, I., Skirbekk, V., & Prskawetz, A. (2010). Social involvement, behavioural risks and cognitive functioning among older people. *Ageing & Society, 30*(5), 779–809.
Fadayevatan, R., Bahrami, M., Mohamadzadeh, M., & Borhaninejad, V. (2020). Relationship of sleep quality with mental health and blood sugar control in elderly people with diabetes mellitus. *Iranian Journal of Ageing, 14*(4), 380–391.

Feng, Z., Falkingham, J., Liu, X., & Vlachantoni, A. (2017). Changes in living arrangements and mortality among older people in China. *SSM-Population Health, 3*, 9–19.

Fernandez-Rodrigues, V., Sanchez-Carro, Y., & Lagunas, L. (2022). Risk factors for suicidal behaviour in late-life depression:A systematic review. *World Psychiatry, 1*, 187–203.

Fülöp, T., Larbi, A., Dupuis, G., & Pawelec, G. (2017). Ageing, frailty, and age-related diseases. *Biogerontology, 18*(6), 721–723.

Gan, T., Yang, J., Jiang, L., & Gao, Y. (2021). Living alone and cardiovascular outcomes: A meta-analysis of 11 cohort studies. *Psychology Health & Medicine, 31*, 1–13.

Gazerro, M., Inelmen, E., Secco, G., & Gatto, M. (1996). Elderly people: State of health and living environment. The Case of Budrio (Northern Italy). *Health & Place, 2*(2), 115–123.

Gezer, C., Yurt, M., Harmancioğlu, B., Yildirim, A., & Elmas, C. (2020). The relationship between malnutrition, diet quality and health-related quality of life among the elderly: A cross-sectional study. *Progress in Nutrition, 22*, e2020065.

Giles, J., Wang, D., & Zhao, C. (2011). Can China's rural elderly count on support from adult children? Implications of rural-to-urban migration. *Journal of Population Ageing, 3*(3), 183–189.

Glaser, K., di Gessa, G., & Tinker, A. (2014). *Grandparenting in Europe. The health and wellbeing of grandparents caring for grandchildren: The Role of cumulative advantage/disadvantage* (Economic and Social Research Council Report, Issue).

Glei, D. A., Goldman, N., Liu, I. W., & Weinstein, M. (2013). Sex differences in trajectories of depressive symptoms among older Taiwanese: The contribution of selected stressors and social factors. *Aging and Mental Health, 17*(6), 773–783.

Grando, V., & Grando, R. (2020). Teaching family nurse practitioner students to provide mental health care to older adults. *Innovation in Aging, 4*(Supplement_1), 10–11.

Gu, D., & Feng, Q. (2018). Psychological resilience of Chinese centenarians and its associations with survival and health: A fuzzy set analysis. *The Journals of Gerontology. Series B, Psychological Sciences and Social Sciences, 73*(5), 580–589. https://doi.org/10.1093/geronb/gbw071

Guo, P., & Li, K. (2012). Trends in esophageal cancer mortality in China during 1987–2009: Age, period and birth cohort analyzes. *Cancer Epidemiology, 12*, 003–105.

Guo, H., Lin, K., Yang, K., Ma, Z., Cao, M., Hu, Y., & Yan, Y. (2022). Trends of cancer incidence among Chinese older adults from 2005 to 2016: A log-linear regression and age-period-cohort analysis. *Frontiers in Public Health*, 1023276.

Hao, L., Xu, X., & Dupre, M. E. (2020). Adequate access to healthcare and added life expectancy among older adults in China. *BMC Geriatrics, 20*, 129. https://doi.org/10.1186/s12877-020-01524-9

Hawkley, L., Wroblewski, K., Kaiser, T., Luhmann, M., & Schumm, L. (2019). Are U.S. older adults getting lonelier? Age, period, and cohort differences. *Psychology and Aging, 34*(8), 1144–1157.

Heo, J., Chun, S., Lee, S., Lee, K. H., & Kim, J. (2015). Internet use and well-being in older adults. *Cyberpsychology, Behavior, and Social Networking, 18*, 268–272.

Hilbrand, S., Coall, A. D., Gerstorf, D., & Hertwig, R. (2017a). Caregiving within and beyond the family is associated with lower mortality for the caregiver. *Evolution and Human Behavior, 38*(3), 397–403.

Hilbrand, S., Coall, D. A., Meyer, A. H., Gerstorf, D., & Hertwig, R. A. (2017b). Prospective study of associations among helping, health, and longevity. *Social Science & Medcine, 187*, 109–117.

Hu, P., Gong, B., & Fan, W. (2017). Influencing factors on mental health of the elderly under network environment. In *2017 International conference on sports, arts, education and management engineering*. Atlantis Press.

Hu, X., Gu, S., Zhen, X., Sun, X., Gu, Y., & Dong, H. (2021). Trends in cognitive function among Chinese elderly From 1998 to 2018: An age-period-cohort analysis. *Frontiers in Public Health*, 753871.

Hu, X., Gu, S., Zhen, X., Sun, X., Gu, Y., & Dong, H. (2022). Trends in activities of daily living disability among Chinese older adults from 1998 to 2018: An age-period-cohort analysis. *European Journal of Ageing, 19*(4), 1167–1179.

Hughes, M. E., & Waite, L. J. (2002). Health in household context: Living arrangements and health in Late Middle Age. *Journal of Health and Social Behavior, 43*(1), 1–21.

Hughes, M. E., Waite, L. J., LaPierre, T. A., & Luo, Y. (2007). All in the family: The impact of caring for grandchildren on grandparents' health. *The Journal of Gerontology Series B: Psychological Science and Social Science, 62*, S108–S119.

Inoue, Y., Wann, D. L., Lock, D., Sato, M., Moore, C., & Funk, D. C. (2020). Enhancing older adults' sense of belonging and subjective well-being through sport game attendance, team identification, and emotional support. *Journal of Aging and Health, 32*(7–8), 530–542.

Jang, H., & Tang, F. Y. (2016). Effects of social support and volunteering on depression among grandparents raising grandchildren. *The International Journal of Aging, 83*(4), 491–507.

Jerger, J., Jerger, S., & Pirozzolo, F. (1991). Correlational analysis of speech audiometric scores, hearing loss, age, and cognitive abilities in the elderly. *Ear and Hearing, 12*(2), 103–109.

Jergovic, M., Coplen, C., Uhrlaub, J., & Nikolich-Zugich, J. (2021). Immune response to COVID-19 in older adults. *The Journal of Heart and Lung Transplantation, 4*(10), 1082–1089.

Joe, J., & Demiris, G. (2013). Older adults and mobile phones for health: A review. *Journal of Biomedical Informatics, 46*, 947–954.

John, A., & Philip, K. (2013). Risk of high dietary calcium for arterial calcification in older adults. *Nutrients, 5*(10), 3964–3974.

Jolie, C., & Lam, B. (2006). Digital inclusiveness–Longitudinal study of internet adoption by older adults. *Journal of Management Information Systems, 22*(4), 177–206.

Kawachi, I., Daniels, N., & Robinson, D. E. (2005). Health disparities by race and class: Why both matter. *Health Affairs, 24*(2), 343–352.

Kim, I. K., & Kim, C. S. (2003). Patterns of family support and the quality of life of the elderly. *Social Indicators Research, 62*(1), 437–454.

Kim, S., Symons, M., & Popkin, B. (2004). Contrasting socioeconomic profiles related to healthier lifestyles in China and the United States. *American Journal of Epidemiology, 159*(2), 184–191.

Kim, O., Byeon, Y., Kim, J., Endo, E., Akahoshi, M., & Ogasawara, H. (2009). Loneliness, depression and health status of the institutionalized elderly in Korea and Japan. *Asian Nursing Research, 3*(2), 63–70.

Kim, J., Choi, Y., Choi, J. W., Nam, J. Y., & Park, E.-C. (2017). Impact of family characteristics by marital status of cohabitating adult children on depression among Korean older adults. *Geriatric Gerontology International, 17*(12), 2527–2536.

Kim, H., Park, H., Lee, H., & Lee, Y. (2019). Hemoglobin levels and low bone mineral density in non-anemic older adults: Secondary analysis of the Korean National Health and Nutrition Examination Survey. *Experimental Gerontology, 126*, 110706–110712.

Krause, N., & Liang, J. (1993). Stress, social support, and psychological distress among the Chinese elderly. *Journal of Gerontology, 48*(6), P282–P291.

Ku, L.-J., Stearns, S., Van Houtven, C., Lee, S.-Y., Dilworth-Anderson, P., & Konrad, T. (2013). Impact of Caring for Grandchildren on the Health of Grandparents in Taiwan. *The Journals of Gerontology: Series B, 68*, 1009–1021.

Laaksonen, M., Prättälä, R., & Lahelma, E. (2002). Sociodemographic determinants of multiple unhealthy behaviors. *Scandinavian Journal of Public Health, 30*, 1–7.

Lafortune, L., Béland, F., Bergman, H., & Ankri, J. (2009). Health state profiles and service utilization in community-living elderly. *Medical Care, 47*(3), 286–294.

Lampek, K., Csóka, L., Hegedüs, R., Zrínyi, M., & Törőcsik, M. (2021). Sports activities of 60 above Hungarian elderly-explaining and predicting impact of exercise on health. *BMC Public Health, 21*, 1863.

Lawton, M. P., Moss, M., & Kleban, M. H. (1984). Marital status, living arrangements and the well-being of older people. *Research on Aging, 6*(3), 323–345.

Lee, R. Y. W., & Carlisle, A. J. (2011). Detection of falls using accelerometers and mobile phone technology. *Age and Ageing, 40*, 690–696.

Lee, Y. H., Chang, Y. C., Chiang, T., Liu, C. T., & Shelley, M. (2019). Living arrangements and sleep-related outcomes among older adults in China: A panel analytic approach. *The International Journal of Aging and Human Development*, Published online.

Leung, J. C. B. (2010). Residential care services for older people in China: From state to market provisions? *Social Development Issues, 32*, 31–47.

Levitt, M. J., Guacci, N., & Weber, R. A. (1992). Intergenerational support, relationship quality, and well-being: A bicultural analysis. *Journal of Family Issues, 13*(4), 465–481.

Li, J. (2012). *Research on culture of contemporary chinese urban elders*. Shanghai People's Press.

Li, L., Li, S., & Zhang, W. (2006). Intergenerational support and self-rated health of rural elderly. *Chinese Gerontology Studies (Zhongguo Laonianxue Zazhi), 26*(11), 1453–1455.

Li, A., Gao, R., Zheng, Q., & Gao, Q. (2017a). Advance in exercise prescription to improve muscular strength and cardiorespiratory function for older adult (review). *Chinese Journal of Rehabilitation Theory and Practice (in Chinese), 2*, 1611–1615.

Li, M., Wan, X., & Wang, Y. (2017b). Time trends of esophageal and gastric cancer mortality in China, 1991–2009: An age-period-cohort analysis. *Scientific Reports, 7*(1), 6797.

Li, Z., Zhang, X., & Lv, Y. (2020). Leisure activities and all cause mortality among the Chinese oldest-old population: A prospective community-based cohort study. *Journal of the American Medical Directors Association, 21*(6), 713–719.e712. https://doi.org/10.1016/j.jamda.2019.08.003

Li, S., Fan, W., Zhu, B., Ma, C., Tan, X., & Gu, Y. (2022). Efects of age, period, and cohort on the prevalence of frailty in Chinese older adults from 2002 to 2014. *Frontiers in Public Health*, 935163.

Liang, Y., Welmer, A., Möller, J., & Qiu, C. (2017). Trends in disability of instrumental activities of daily living among older Chinese adults, 1997–2006: Population based study. *BMJ Open, 7*(8), e016996.

Lin, S., Beck, A., Finch, B., Hummer, R., & Masters, R. (2012a). Trends in US older adult disability: Exploring age, period, and cohort effects. *American Journal of Public Health, 102*(11), 2157–2163.

Lin, S., Beck, A., Finch, B., Hummer, R., & Master, R. (2012b). Trends in US older adult disability: Exploring age, period, and cohort effects. *American Journal of Public Health, 103*(1), e8.

Lin, C., Wang, F., Lin, S., Chen, C., & Huang, C. (2014a). A comparison of inpatients with anxious depression to those with nonanxious depression. *Psychiatry Research, 220*(3), 855–860.

Lin, S., Beck, A., & Finch, B. (2014b). Black-white disparity in disability among U.S. older adults: Age, period, and cohort trends. *The Journals of Gerontology. Series B. Psychological and Social Sciences, 69*(5), 784–797.

Liu, X., Chen, J., Zhou, J., Liu, J., Lertpitakpong, C., Tan, A., & Mao, Z. (2019). The relationship between the number of daily health-related behavioral risk factors and sleep health of the elderly in China. *International Journal of Environmental Research and Public Health, 16*(24), 4905.

Liu, X., Yu, Y., Wang, M., & Mubarik, S. (2020). The mortality of lung cancer attributable to smoking among adults in China and the United States during 1990–2017. *Cancer Communications, 40*(11), 611–619.

Lu, W., Pai, M., Scholes, S., & Xue, B. (2021). Do depressive symptoms link chronic diseases to cognition among older adults? Evidence from the health and retirement study in the United States. *Journal of Affective Disorders, 294*, 357–365.

Lund, B. D., & Wang, T. (2020). A bibliometric study of research pertaining to the oldest-old (age eighty-five and older). *Journal of the Medical Library Association, 108*(1), 1–16. https://doi.org/10.5195/jmla.2020.762

Lund, R., Due, P., Modving, J., Holstein, B. E., Damsgaard, M. T., & Anderson, P. K. (2002). Cohabitation and marital status as predictors of mortality. *Social Science & Medicine, 55*, 673–679.

Luo, Y., & Waite, L. J. (2005). The impact of childhood and adult SES on physical, mental, and cognitive well-being in later life. *The Journals of Gerontology Series B: Psychological Sciences & Social Sciences, 60*(2), S93–S101.

Lv, M., & Zhang, Y. (2021). Does internet use help elderly people get intergenerational support?—Heterogeneity research based on urban and rural dimensions. *The World of Survey and Research (In Chinese), 7*, 35–45.

Lynch, S. M. (2003). Cohort and life-course patterns in the relationship between education and health: A hierarchical approach. *Demography, 40*(2), 309–331.

Magaziner, J., Cadigan, D. A., Hebel, J. R., & Parry, R. E. (1988). Health and living arrangements among older women: Does living alone increase the risk of illness? *Journal of Gerontology, 43*(5), M127–M133.

Marttila, E., Koivula, A., & Rasanen, P. (2021). Does excessive social media use decrease subjective well-being? A longitudinal analysis of the relationship between problematic use, loneliness and life satisfaction. *Telematics and Informatics, 59*, 101556.

Michael, Y. L., Berkman, L. F., Gold, G. A., & Kawachi, I. (2001). Living arrangements, social integration and change in functional health status. *American Journal of Epidemiology, 153*(2), 123–131.

Muusses, L. D., Finkenauer, C., Kerkhof, P., & Billedo, C. J. (2014). A longitudinal study of the association between compulsive Internet use and wellbeing. *Computers in Human Behavior, 36*, 21–28.

Nabi, R. L., Prestin, A., & So, J. (2013). Facebook friends with (health) benefits? Exploring social network site use and perceptions of social support, stress, and well-being. *Cyberpsychology, Behavior and Social Networking, 16*, 721–727.

Oshio, T. (2022). Is caring for grandchildren good for grandparents' health? Evidence from a fourteen-wave nationwide survey in Japan. *Journal of Epidemiology, Published online, 32*, 363–369.

Pampel, F. C., Krueger, P. M., & Denney, J. T. (2010). Socioeconomic disparities in health behaviors. *Annual Review of Sociology, 36*, 349–370.

Pan, C., Cao, N., Kelifa, M., & Luo, S. (2023). Age and cohort trends in disability among Chinese older adults. *Frontiers in Public Health, 11*, 998948.

Peng, Y., & Chan, Y. (2020). Do internet users lead a healthier lifestyle? *Journal of Applied Gerontology, 39*(3), 277–284.

Perng, A., & Renz, S. (2017). Identifying and treating complicated grief in older adults. *The Journal for Nurse Practitioners, 14*(4), 289–295.

Petrovcic, A., Fortunati, L., Vehovar, V., Kavcic, M., & Dolnicar, V. (2015). Mobile phone communication in social support networks of older adults in Slovenia. *Telematics and Informatics, 32*(4), 642–655.

Plaza, I., Martín, L., Martin, S., & Medrano, C. (2011). Mobile applications in an aging society: Status and trends. *The Journal of Systems and Software, 84*, 1977–1988.

Reiland, H., Giuca, A., Banerjee, A., Hillard, C., & Goveas, J. (2021). The relationship of loneliness with mental health trajectories in bereaved older adults. *The American Journal of Geriatric Psychiatry, 29*(4), S88–S89.

Ren, N., Li, Y., & Wan, Z. (2022). Patterns of HIV or AIDS mortality among older people from 1990 to 2019 in China: Age-period-cohort analysis. *JMIR Public Health and Surveillance, 8*(11), e35785.

Rendina, H., Millar, B., Gurung, S., Starks, T., & Parsons, J. (2017). Psychsocial wellbeing and immune health among HIV-positive older adults: The critical roel of HIV stigma. *Psychosomatic Medicine: Journal of the American Psychosomatic Society, 79*(4), A102–A103.

Rook, K. S. (1984). The negative side of social interaction: Impact on psychological well-being. *Journal of Gerontology, 46*(5), 1097–1108.

Rook, K. S., & Pietromonaco, P. (1987). Close relationship: Ties that heal or ties that bind? *Advances in Personal Relationships, 1*, 1–35.

Sereny, M. D., & Gu, D. (2011). Living arrangement concordance and its association with self-rated health among institutionalized and community-residing older adults in China. *Journal of Cross Culture Gerontology, 26*, 239–259.

Shapiro, J., Kuo, H., Morgan, R., Li, H., & Leng, S. (2020). Intersection of sex and frailty in humoral immune responses to influenza vaccine in community-dwelling older adults. *Innovation in Aging, 4*(S1), 271–272.

Shi, Z., Zhang, T., & Byles, J. (2015). Food habits, lifestyle factors and mortality among oldest old Chinese: The Chinese longitudinal healthy longevity survey (CLHLS). *Nutrients, 7*(9), 7562–7579.

Siira, H., Elo, S., Kyngäs, H., & Falck, A. (2022). Well-being supportive home environment of elderly people with visual impairments and health-related quality of life. *Activities, Adaptation & Aging, 46*(3), 251–267.

Song, B., Yoshino, H., Goto, T., Zhang, H., Guan, J., Hasegawa, K., & Zhang, Q. (2022). Indoor thermal environment during heating season and the health of elderly in China. *Science and Technology for the Built Environment, 28*(7), 843–863.

Strand, B., Bergland, A., Jørgensen, L., Schirmer, H., Emaus, N., & Cooper, R. (2019). Do more recent born generations of older adults have stronger grip? A comparison of three cohorts of 66- to 84-year-olds in the Tromsø Study. *The Journals of Gerontology. Series A, Biological Sciences and Medical Sciences, 74*(4), 528–533.

Sullivan, K., Liu, R., Dodge, H., & Andreescu, C. (2019). Depressive symptoms on the decline in older adults: Birth cohort analysis from the rust belt. *American Journal of Geriatric Psychiatry, 01*, 023.

Sun, Y., Wang, Y., & Li, M. (2018). Long-term trends of liver cancer mortality by gender in urban and rural areas in China: An age-period-cohort analysis. *BMJ Open, 8*(2), e020490.

Suzman, R., Willis, D. P., & Manton, K. G. (1992). *The oldest-old.* Oxford University Press.

Swickert, R., & Hittner, J. (2002). Relationship among Internet use, personality, and social support of the elderly. *Computers in Human Behavior, 18*(4), 437–451.

Tamai, K., Terai, H., Takahashi, S., Katsuda, H., Shimada, N., Habibi, H., & Nakamura, H. (2022). Decreased daily exercise since the COVID-19 pandemic and the deterioration of health-related quality of life in the elderly population: A population-based cross-sectional study. *BMC Geriatrics, 22*(1), 678.

Thalmann, A., Morfeld, M., & Benthien, A. (2010). Anxiety and depression in the elderly: Results of a regional examination. *Gesundheitswesen, 74*(1), 29–33.

Tian, B. H., & Wang, C. Y. (2014). Structural characteristics and determinants of the demand of the elderly for social care services-based on the analysis of the replacement mechanism of family supporting for the elderly. *China Rural Survey (in Chinese), 4*, 2–17.

Tsai, F.-J., Motamed, S., & Rougemont, A. (2013). The protective effect of taking care of grandchildren on elders' mental health? Associations between changing patterns of intergenerational exchanges and the reduction of elders' loneliness and depression between 1993 and 2007 in Taiwan. *BMC Public Health, 13*, 567.

Tsai, H. Y. S., Shillair, R., Cotten, S., Winstead, V., & Yost, E. (2015). Getting grandma online: Are tablets the answer for increasing digital inclusion for older adults in the U.S.? *Educational Gerontology, 41*, 695–709.

Tylavsky, F., Anderson, J., & Talmage, R. (1985). The effects of calcium intake and estrogen, dituretic, and antacid use on bone health in elderly females. *American Journal of Epidemiology, 122*(3), 525–526.

Veena, C., Kwon, W., & Juan, G. (2012). Internet use and perceived impact on quality of life among older adults: A phenomenological investigation. *International Journal of Health, 2*(3), 1–13.

Waldrop, D. P., & Weber, J. A. (2001). From grandparent to caregiver: The stress and satisfaction of raising grandchildren. *Families in Society, 82*, 461–472.

Wei, J., Ali, M., Wang, T., & Xu, H. (2021a). Abdominal aortic calcification and cognitive function among older adults: Cross-sectional analysis of National Health and Nutrition Examination Survey, 2013–2014. *International Journal of Geriatric Psychiatry, 36*(11), 1778–1784.

Wei, J., Hou, R., Xie, L., & Chandrasekar, E. (2021b). Sleep, sedentary activity, physical activity, and cognitive function among older adults: The National Health and Nutrition Examination Survey, 2011–2014. *Journal of Science and Medicine in Sport, 24*(2), 189–194.

Wei, K., Liu, Y., Yang, J., Gu, N., Cao, X., Zhao, X., Jiang, L., & Li, C. (2022). Living arrangement modifies the associations of loneliness with adverse health outcomes in older adults: Evidence from the CLHLS. *BMC Geriatrics, 22*, 59.

Wei, L., Li, Y., & Zhu, Y. (2023). Age-period-cohort study of Chinese older adults' suicide. *Chinese Journal of Geriatric Studies (in Chinese), 6*, 1486–1490.

Wen, M., & Gu, D. (2021). Living arrangements and disability among older adults in China. *China: An International Journal, 19*(3), 132–147. https://doi.org/10.1353/chn.2021.0032

Wu, Q., & Gu, D. (2021). Oldest-old adults. In D. Gu & D. M. Dupre (Eds.), *Encyclopedia of gerontology and population aging* (pp. 3637–3653). Springer Publisher.

Xiang, X., & Cheng, J. (2019). Trajectories of major depression in middle-aged and older adults: A population-based study. *International Journal of Geriatric Psychiatry, 34*(10), 1506–1514.

Xu, L., Wu, B., Chi, I., & H-Y., H. (2012). Intensity of grandparent caregiving and life satisfaction among rural Chinese older adults: A longitudinal study using latent difference score analysis. *Family & Community Health, 35*, 287–299.

Xu, R., Liu, Y., Mu, T., Ye, Y., & Xu, C. (2022). Determining the association between diferent living arrangements and depressive symptoms among over-65-year-old people: The moderating role of outdoor activities. *Frontiers in Public Health, 10*, 954416.

Yang, G., & D'Arcy, C. (2023). Age, period and cohort effects in depression prevalence among Canadians 65+, 1994 to 2018: A multi-level analysis. *International Journal of Social Psychiatry, 69*(4), 885–894.

Yu, R. P., Mccammon, R. J., Ellison, N. B., & Langa, K. M. (2016a). The relationships that matter: Social network site use and social wellbeing among older adults in the United States of America. *Ageing & Society, 36*, 1826–1852.

Yu, R., Wong, M., Chang, B., Lai, X., Lum, C., Auyeung, T., Lee, J., Tsoi, K., Lee, R., & Woo, J. (2016b). Trends in activities of daily living disability in a large sample of community-dwelling Chinese older adults in Hong Kong: An age-period-cohort analysis. *BMJ Open, 6*(12), e013259.

Yu, R., Wong, W., Chong, K., Chang, B., Lum, C., Auyeung, T., Lee, J., Lee, R., & Woo, J. (2018a). Trajectories of frailty among Chinese older people in Hong Kong between 2001 and 2012: An age-period-cohort analysis. *Age & Aging, 47*(2), 254–261.

Yu, Y., Ma, L., Lei, J., Li, J., & Ma, C. (2018b). Relationship between community support and cognitive function in the elderly: Data analysis from the China longitudinal study on elderly health factors. *Chinese Journal of Mental Health, 6*, 490–494.

Zeng, Y., Cheng, L., & Chen, H. (2010). Effects of FOXO genotypes on longevity: A biodemographic analysis. *The Journals of Gerontology. Series A, Biological Sciences and Medical Sciences, 65*(12), 1285–1299. https://doi.org/10.1093/gerona/glq156

Zeng, Y., Gu, D., & George, L. K. (2011). Association of religious participation with mortality among Chinese old adults. *Research on Aging, 33*, 51–83. https://doi.org/10.1177/0164027510383584

Zeng, Y., Chen, H., & Ni, T. (2015). GxE interactions between FOXO genotypes and tea drinking are significantly associated with cognitive disability at advanced ages in China. *Journals of Gerontology Series A: Biological Sciences and Medical Sciences, 70*(4), 426–433. https://doi.org/10.1093/gerona/glu060

Zeng, Y., Feng, Q., & Hesketh, T. (2017). Survival, disabilities in activities of daily living, and physical and cognitive functioning among the oldest-old in China: A cohort study. *Lancet, 389*(10079), 1619–1629. https://doi.org/10.1016/s0140-6736(17)30548-2

Zeng, Y., Nie, C., & Min, J. (2018). Sex differences in genetic associations with longevity. *JAMA Network Open, 1*(4), e181670. https://doi.org/10.1001/jamanetworkopen.2018.1670

Zhang, W. (2008). Religious participation and mortality risk among the oldest old in China. *Journals of Gerontology – Series B Psychological Sciences and Social Sciences, 63*(5), S293–S297. https://doi.org/10.1093/geronb/63.5.s293

Zhang, L. (2021). A latent class analysis of health lifestyles and health outcomes among Chinese older adults. *Ageing & Society, 41*, 313–338.

Zhang, Q., & Li, Z. (2022). The impact of internet use on the social networks of the elderly in China—The mediating effect of social participation. *International Journal of Environmental Research and Public Health, 19*, 9576.

Zhao, Y., Song, J., Brytek-Matera, A., Zhang, H., & He, J. (2021). The relationships between sleep and mental and physical health of Chinese elderly: exploring the mediating roles of diet and physical activity. *Nutrients, 13*(4), 1316.

Zhou, Z., Mao, F., Ma, J., Hao, S., Qian, Z., Elder, K., Turner, J., & Fang, Y. (2018). A longitudinal analysis of the association between living arrangements and health among older adults in China. *Research on Aging, 40*(1), 72–97.

Zhou, N., Wen, J., Stelzer, E., & Killikelly, C. (2020). Prevalence and associated factors of prolonged grief disorder in Chinese parents bereaved by losing their only child. *Psychiatry Research, 284*, 112766.

Zhu, H., & Xie, Y. (2007). Socioeconomic differentials in mortality among the oldest old in China. *Research on Aging, 29*, 125–143. https://doi.org/10.1177/016402750629

Zhu, Y., Guo, X., Zhang, X., Shi, X., Yang, Y., & Zhang, Q. (2022). Sex differences in the relationship of serum creatinine to cystatin C ratio and depressive symptoms among middle-aged and older adults in China. *Journal of Affective Disorders, 319*, 57–61.

Zimmer, Z. (2005). Health and living arrangement transitions among Chinese oldest-old. *Research on Aging, 27*(5), 526–555.

Zimmer, Z., Chayovan, N., Lin, H.-S., & Natividad, J. (2004). How indicators of socioeconomic status relate to physical functioning of older adults in three asian societies. *Research on Aging, 26*(2), 224–258.

Zimmer, Z., Martin, L. G., & Nagin, D. S. (2012). Modeling disability trajectories and mortality of the oldest-old in China. *Demography, 49*(1), 291–314. https://doi.org/10.1007/s13524-011-0075-7

Zivin, K., Pirraglia, P., McCammon, R., Langa, K., & Vijan, S. (2013). Trends in depressive symptom burden among older adults in the United States from 1998 to 2008. *Journal of General Internal Medicine, 28*(12), 1611–1619.

Zunzunegui, M., Beland, F., & Otero, A. (2001). Support from children, living arrangements, self-rated health and depressive symptoms of older people in Spain. *International Journal of Epidemiology, 30*, 1090–1099.

Data and Methods

3

Since the results and conclusions of each analysis can be affected by data and research methodologies involved, this chapter devotes to an explication of data and methods used in this study. The chapter first introduces datasets that are used in this manuscript. It then introduces several other datasets that can be possibly used in demographic and sociological studies of Chinese older adults' (including the oldest old's) health from a longitudinal point of view. In the methods section, the chapter introduces the age-period-cohort approach and the cross-classified random-effects models (CCREMs). Such discussion may enlighten readers the data, methodological and statistical issues that I have confronted and solved in the analyses. A detailed discussion on data and methods in this chapter also allows future research to both replicate and extend findings of this research.

3.1 Data

Several datasets that surveyed Chinese samples, especially older adults, including the oldest old, became available in recent decades, which can be considered as breakthroughs in studying older adults in China. In this part of the chapter, I will explicate available data sources, including the one that is analyzed in this study. All these datasets have special features, research interests and goals. And all of them are designed at the individual level, which contain extremely detailed micro-level data that allow researchers to explore health of older adults, including health the oldest old, and the influential factors of health at the individual level. The following subsections will describe features of these datasets and point out why they are suitable for longitudinal studies of older adults' (including the oldest old's) health.

L. Zhang, *Living Longer and Healthier at Older Ages*, Advances in Studies of Aging and Health 3, https://doi.org/10.1007/978-3-031-69773-9_3

3.1.1 The Chinese Longitudinal Healthy Longevity Survey (CLHLS)

The first data source that will be depicted here is the Chinese Longitudinal Healthy Longevity Survey (CLHLS). The CLHLS data have been collected by Peking University's Center for Healthy Aging and Family Studies and the China National Research Center on Aging, with support from the U.S. National Institute on Aging. The CLHLS is the first nationwide longitudinal survey on the determinants of healthy longevity. It is considered as "the first large surveys of the oldest old conducted in a developing country" (Yi Zeng et al., 2002: 252).

The survey was initiated in 1998 and nearly 9000 Chinese older adults were recruited from more than 800 randomly selected counties and cities (Liaoning, Jilin, Heilongjiang, Beijing, Tianjin, Shanxi, Shaanxi, Shanghai, Jiangsu, Zhejiang, Anhui, Fujian, Jiangxi, Shandong, Henan, Hubei, Hunan, Guangdong, Guangxi, Sichuan, Chongqing and Hainan). The survey covered roughly half the counties and cities of those provinces and the sample areas, which represented 85.0% of the total population of China.

Until now, eight waves (1998, 2000, 2002, 2005, 2008, 2011–12, 2014 and 2018) of survey data have been collected, which allows researchers to conduct longitudinal analyses to investigate changing patterns of older adults' health, especially the oldest old's health status and a variety of issues. The samples were followed for every 2–3 years until death or contact was lost. In each wave, a large number of new samples were added to replace the lost respondents. In the 2011–2012 survey, for example, 3,802 male respondents and 4,603 female respondents (for a total of 8,405) surveyed in 2008 were re-interviewed. And 801 males and 982 females (for a total of 1,783) were newly added interviewees. A similar rationale applies to other survey waves as well.

During the past two decades, the survey has conducted nearly 113,000 interviews from more than 56,700 participants with the majority of them being 80 years or older. All interviews were face-to-face with informed consent. In total, the CLHLS collected data for about 29,000 deceased persons. Given that the survey was initially launched to meet the needs for scientific research on the oldest old, the dataset provides an excellent source for studying the oldest old in China. Meanwhile, considering the longitudinal design of CLHLS surveys, this dataset also serves as an excellent source for studying health of Chinese older adults (including the oldest old) from a longitudinal point of view. Thus, empirical studies of this manuscript are all based on analyzing this data source.

It was pointed out that persons who reported age 106 or higher were considered as invalid cases (Zeng et al., 2002). Thus, when analyzing this dataset, persons aged 106 and higher are not included in this study due to insufficient information to validate their reported extremely high age. The samples studied in this research are those aged 85 to 105. This manuscript relies on analyzing eight waves of the CLHLS data spanning 20 years to elucidate changing patterns of Chinese oldest old's health status as well as its demographic, socioeconomic and behavioral determinants. The results show how health outcomes of Chinese oldest old have changed by age, period and cohort when a variety of demographic, socioeconomic and behavioral factors are controlled. Detailed findings will be shown in following chapters of this book. Besides the CLHLS data, I also introduce several other available datasets that allow researchers to conduct longitudinal studies of Chinese older adults' health.

3.1.2 The China Family Panel Studies (CFPS)

The second dataset is China Family Panel Studies (CFPS), which is a nationally representative, annual longitudinal survey of Chinese communities, families and individuals. The CFPS is funded by the Chinese government through Peking University. The survey was launched in 2010 by the Institute of Social Science Survey (ISSS) at Peking University of China. The CFPS is designed to collect individual-, family-, and community-level longitudinal data on wellbeing of the Chinese population. Information collected by CFPS deals with a wide range of topics, including individual economic activities, education outcomes, family dynamics and relationships, migration, health et al.

CFPS has three key features: First, all members above age 9 in a sampled household are interviewed and they constitute core members of the CFPS. Second, children of the CFPS are also considered core members of the CFPS. A core member is considered as leaving the study only through death. Third, all core members are re-interviewed on a yearly basis. Five provinces were chosen for initial oversampling (1,600 families in each). By doing so, regional comparisons become possible. The remainder of the CFPS samples (8,000 families) were drawn from other provinces so that the overall CFPS samples well represent population of the country.

The 2010 CFPS survey is considered as the baseline survey, which used multi-stage probability strategy to obtain samples. The CFPS samples have been collected through three stages: county (or equivalent), village (or equivalent) and household. Interviews have been conducted using computer assisted personal interviewing (CAPI) technology, provided by the Survey Research Center (SRC) at the University of Michigan. The 2010 baseline survey has interviewed almost 15,000 families and almost 30,000 individuals within these families for an approximately response rate of 79.0% (http://www.isss.pku.edu.cn/cfps/en/index.htm). The three waves of CFPS data allow researchers to examine health related issues among Chinese older adults from a longitudinal point of view.

3.1.3 The Chinese General Social Survey (CGSS)

The third data source to be introduced here is the Chinese General Social Survey (CGSS), which is considered as the earliest national longitudinal survey conducted by academic institutions in mainland China. The Department of Social Sciences at Renmin University in China and Center of the Hong Kong University of Science and Technology launched the earliest survey in 2003. The first five-year survey data were collected during 2003 to 2008 (excluding 2007) in the first phase. Seven surveys were then conducted in the second phase (2010–2019, excluding 2014).

The CGSS has applied a multi-stage stratified approach to survey China's urban and rural households. The survey has been conducted annually or biannually to gather longitudinal data. All surveys have employed face-to-face interviews to collect information from the respondents. The survey aims to systematically capture the changing relationship between social structure and quality of life in both urban and rural China. Information relating to social structure includes information on social group, organizations as well as networks of the respondents' social relationships. Quality of life is defined as the objective and subjective of well-being at the individual and aggregate levels (http://cgss.ruc.edu.cn). Some basic information about Chinese residents' behavior and thinking mode, lifestyles, and social changes has been collected as well (Han & Zhao, 2021).

Until now, multiple waves of CGSS data are available for public use. Researchers may take advantage of the dataset's longitudinal nature as well as its information on individual health and wellbeing to study the changing patterns of health among Chinese older adults.

3.1.4 The China Health and Retirement Longitudinal Study (CHARLS)

The next dataset to be introduced here is China Health and Retirement Longitudinal Study (CHARLS) which is a nationally representative longitudinal survey of adults aged 45 and older and their spouses. The design of CHARLS is based on the Health and Retirement Study (HRS) and related aging surveys such as the English Longitudinal Study of Aging (ELSA), the Survey of Health, and Aging and Retirement in Europe (SHARE). CHARLS is conducted by the China Centre for Economic Research of Peking University. The baseline wave of CHARLS was conducted in 2011 which included about 10,000 households and 17,500 individuals in 150 counties/districts and 450 villages/committees. The follow-up surveys were conducted for every 2 years, with wave 2 in 2013, wave 3 in 2015, and wave 4 in 2018 (https://charls.pku.edu.cn/).

The CHARLS questionnaire includes the following modules: (1) demographics; (2) family structure/transfer; (3) health; (4) health insurance and healthcare utilization; (5) work, retirement, and pension; (6) relative income; (7) family income, wealth, and expenditures; (8) personal income and assets; and (9) housing characteristics. In addition to individual level information, community level information is collected as well. Detailed information relating to the above modules has been collected based on face-to-face interviews in respondents' homes.

Using a multi-stage probability-proportional-to-size technique, participants have been randomly sampled by region and then by urban district or rural county and their per capita GDP. In total, CHARLS has sampled residents from 150 counties across 28 provinces in China. According to data on 17,708 individuals in the national baseline sample, 52.1% of them are females. Most of the sample are the young old, 40.0% are aged 60 years and older.

Besides the individual social, economic, and behavioral data, CHARLS also contains rich information on the respondent's health. Such information includes self-reported health, doctors' diagnoses of a set of chronic diseases, eyesight, hearing, and dental health, hedonic well-being, difficulties in activities of daily living (ADLs) and instrumental activities of daily living (IADLs) and physical functioning. In addition, biomarker information is collected as well. The first blood collection of CHARLS was conducted in the baseline survey from 2011 to 2012 by collecting blood samples for 11,847 individuals. The second follow-up wave of blood collection was done in wave 3 (2015) by collecting blood samples for 13,013 individuals. In wave 4, CHARLS (2018) started to have questions collecting information on dementia and mild cognitive impairment among respondents aged 60 and older (Chen et al., 2021).

Thus far, six waves of CHARLS data are available for public use, which are the 2011, 2012, 2013, 2014, 2015 and 2018 waves. The longitudinal design of CHARLS data allow researchers to study factors affecting Chinese older adults' health from a longitudinal scope.

3.1.5 The China Longitudinal Aging Social Survey (CLASS)

An additional dataset is the China Longitudinal Aging Social Survey (CLASS), which is a nationally representative longitudinal survey that was initiated in 2014. By collecting social and economic data of the Chinese aging population regularly and systematically on both individual and community levels, CLASS aims to investigate various problems and challenges that Chinese will face during the aging process, and evaluate the effects of various social security policies on improving life-quality of the older population (http://class.ruc.edu.cn). The CLASS has been collecting data biannually since then. Thus far, three waves of data (2014, 2016, and 2018) are available for public use. The survey has adopted a multistage stratified random sampling procedure with counties as primary sampling units and villages in rural areas and communities in urban areas as secondary sampling units (Li & Zhang, 2023). Thus far, 134 counties have been selected from 30 of China's 34 provinces and municipalities as the primary sampling unit (PSU). From these counties, 462 communities and villages were randomly chosen as the secondary sampling unit (SSU). In the last stage, simple random sampling was utilized to select 25 respondents aged 60 years or above from each SSU. These individuals constitute overall samples of the survey (Yang et al., 2021). The final samples included 462 villages and communities within 134 counties from 28 provinces or municipalities in mainland China (Tibet, Xinjiang and Hainan were excluded due to execution limitations).

This survey has collected information on both parents and their children, including detailed information on economic and housework exchanges between parents and children. Children's information was only reported by their parents. As far as information on older adults, CLASS collected detailed information on older adults' health outcomes, including consistent measures of depressive symptoms spanning 4 years. The data also contained information on functional exchanges and emotional affinity between older parents and each of their adult children. Therefore, this data source may be used to explore Chinese older adults' health by applying a longitudinal approach.

3.1.6 The China Health and Nutrition Survey (CHNS)

The last dataset to be introduced is the China Health and Nutrition Survey (CHNS) which is also a superb data source studying health related issues among Chinese. The CHNS is an ongoing, open-cohort, international collaborative project between the Carolina Population Center at the University of North Carolina at Chapel Hill (UNC-CH) and the National Institute for Nutrition and Health (NINH) at the Chinese Center for Disease Control and Prevention (CCDC) (https://www.cpc.unc.edu/projects/china). Survey protocols, instruments, and the process for obtaining informed consent for CHNS have been approved by the institutional review committees of the UNC-CH, the National Institute of Nutrition and Food Safety, and the CCDC. The data were collected by trained and certified health workers.

In 1979, China started the open-door policy and initiated social and economic reforms. Consequently, the country's economy and agricultural productivity changed greatly and remarkable shifts have occurred in people's diet, activity, and body composition. In such a social background, a large national survey is warranted to examine how economic and social changes have affected a range of health behaviors in this large country. Thus, the first round of the CHNS was conducted in 1989 and has been followed up for every 2–4 years. Since 1989, the CHNS has collected individual, household, and community level data to assess changes in health, nutrition, and family planning behaviors and outcomes. So far, there have been a total of 10 waves of surveys, conducted in 1989, 1991, 1993, 1997, 2000, 2004, 2006, 2009, 2011, and 2015.

A multistage, stratified, sampling design has been applied to select samples from provinces across China. It ensures that the CHNS provides a fair representation of urban and rural areas. Since the 2000 survey, nine provinces across four regions are included: Northeast (Liaoning, Heilongjiang), East Coast (Jiangsu, Shandong), Central (Hennan, Hunan, Hubei) and West (Gunagxi, Guizhou). These provinces represent different levels of socioeconomic development in China. Within each province, counties have been stratified by different income levels and a weighted sampling scheme has been used to randomly select four counties. One county capital town and three villages within the counties and urban/suburban neighborhoods within the cities were subsequently randomly selected. Within each city, 2 urban and 2 suburban neighborhoods were randomly selected; within each county, one community in the capital city and three rural villages were selected. Eventually, twenty households were randomly selected from each village, town or neighborhood. All individuals in each household were interviewed in the CHNS. The surveys provided in-depth measurement over time and in different places in China. A total of 12 social and economic dimensions and physical infrastructures related to health and nutritional status and welfare were included in the surveys (Zhang et al., 2014). As a result, the CHNS has collected data on a total of 388 communities, 11,130 households and 42,829 individuals. The selection of numbers of rural and urban sample was based on the proportion distribution of overall Chinese rural-urban population.

The CHNS aims to document changes in individual health and nutrition, and the associated changes in family planning policies and programs implemented by national and local governments in China. Thus, the in-depth survey collects detailed information on (a) cognitive and physical function (e.g., memory, counting, orientation, activities of daily living (ADL), and instrumental activities of daily living (IADL)); (b) health-related behaviors (e.g., dietary intake, physical activity, time use, and tobacco and alcohol use); (c) mental health and well-being (e.g., stress, sleep, quality of life, and self-rated health); (d) physical health (e.g., anthropometry and blood pressure); and (e) detailed individual, community environment, and sociodemographic data (e.g., urbanization, household size and composition, intergenerational transfers, and assets). Biomarkers from fasting blood, toenails, and genome-wide association data were included in the 2009 and 2015 surveys; microbiome and metabolome related information was included in the 2015 and 2019 surveys (Zhang et al., 2021).

Since the CHNS surveys are not mainly interested in studying older adults, respondents aged 65 and over are only about 10.0% in the 2015 survey. The percentage shares of older adults in other survey years are similar to year 2015. Even with a low percentage share of elder population in the survey, the survey samples cover 11 provinces of China, which well represents older adult profiles in China. Ten waves (1989, 1991,1993, 1997, 2000, 2004, 2006, 2009, 2011, 2015, and 2018–2019) of the dataset allow researchers to apply the APC approach examining older adults' health from a longitudinal scope. In sum, the chapter has introduced several datasets that allow researchers to perform longitudinal analyses of health among older adults, including the oldest old in China. I now move to statistical methods to be employed in this monograph.

3.2 Statistical Methods

3.2.1 Descriptive Analysis

Descriptive methods are used throughout empirical analysis of this book. In Chap. 4 and the following chapters, the study presents basic descriptive information of Chinese oldest old's health status, its determinants. The CCREM results demonstrate significant period and cohort effects in the links between the oldest old's health and a variety of demographic, socioeconomic and behavioral factors. The manuscript shows basic descriptive tables and graphs of data, which show period and cohort variations of Chinese oldest old's health status and its determinants.

Descriptive tables in the following chapters show, for most covariates, the percentage distributions of variables. For continuous variables, such as age, years of schooling, cognitive function score, positive and negative feeling scores, the mean values are shown in descriptive tables. Since CLHLS is a sampling survey and the sampling weight has not been applied, the results refer to the sample.

3.2.2 Regression Methods

Given that the study is interested in examining the period and cohort effects in the association between a variety of demographic, socioeconomic and behavioral factors and Chinese oldest old's health, an age-period-cohort (APC) approach is used in the analysis. Age-period-cohort (APC) models broadly refer to statistical techniques that may detect the effects of chronological age (i.e., age effects), societal conditions at the time of observation (i.e., period effects), and some common experiences shared among individuals during the same historical moment (i.e., cohort effects). Social scientists have applied APC models to understand how age, period, and cohort effects have played a role in shaping individual health outcomes through examining longitudinal or repeated cross-sectional data (Bardo & Reither, 2021).

3.2.2.1 Definitions of Age, Period and Cohort

Age, period, and cohort effects all refer to some types of time-related variations in the phenomena of interest and each dimension carries distinct substantive meanings (Yang, 2008a, b). Age effects are defined as variations associated with different chronological age groups. Such variations are caused by physiological changes, accumulation of social experience, and/or role or status changes (Reither et al., 2009). Thus, age effects reflect biological and social processes of aging internal to individuals and represent developmental changes across the life course. Age is obviously related to one's health status. Health declines when individuals move to old ages.

Period refers to a specific time point in history. A period may be a year or a decade. Period effects are seen as changes among people of all ages from one historical period to another (Schwadel, 2010). Period effects usually occur due to social, cultural and economic changes that begin to take place in certain time periods. These changes induce similar changes in individual health of all ages.

Birth cohort refers to a group of people who were born at the same time. Cohort effects are associated with changes across groups of individuals who experience an initial event such as birth in the same year or years. Cohort effects indicate changes across the birth cohorts regardless of age. These effects represent the effects of formative experience for successive age groups in successive time periods (Glenn, 2003; Ryder, 1965). As Ryder (1965) argued in a classic essay, social change occurs because of individual change or an ongoing massive process of personnel replacement in which older cohorts are succeeded by younger ones with different attitudes and behaviors. This type of population turnover has been called "cohort replacement" or "cohort succession" in subsequent work (Firebaugh, 1989).

The distinction between period effects and cohort effects lies in the way in which people view the causes of social changes. Health outcomes, in particular, can improve or become worse over time due to some relevant historical events experienced by various birth cohorts. This reflects the period effects. On the other hand, health outcomes may get better or worse across successive birth cohorts, which shows the cohort effects. Cohort effects on health can be caused by differentials in early life conditions, which is a commonly cited explanation for susceptibility to diseases and mortality (Barker, 1998). For example, more recent U.S. cohorts are found to have better health capital at birth than their earlier counterparts (Fogel & Costa, 1997). Elder (1987) has also pointed out that the same social change may impact individuals in a significantly different manner depending upon their cohort membership (Elder, 1987).

3.2.2.2 Theoretical Considerations

Ryder (1965) argued that technological innovation is the "principal motor of contemporary social change" (p. 851). Ryder also contended that secular changes such as technological innovation generally influence multiple birth cohorts simultaneously. Indeed, social changes tend to have more important influence on new cohorts because they are more likely than older cohorts to embrace new technologies and their corresponding modes of work and leisure. Older cohorts are more used to past social norms, values and technological standards, which makes them resistant to social changes. In addition, younger birth cohorts are also more likely to benefit from medical improvements. As a result, younger cohorts may have better general health outcomes than older birth cohorts. But at the meantime, individuals in recent birth cohorts have a higher risk of accessing to modern technologies, products and media, which makes them to have a higher likelihood of adopting sedentary lifestyles and consuming high calorie diets. Consequently, they may develop more chronic diseases than their processors.

In light of these theoretical propositions, prior literature scrutinized mortality trends caused by diabetes, liver, esophageal and gastric cancer and documented the age, period and cohort effects (Guo & Li, 2012; Li et al., 2017; Liu et al., 2020; Sun et al., 2018). Taking lung cancer mortality as an example, the results indicated that the lung cancer mortality attributable to smoking was generally increasing among Chinese adults over time. Scholars have also examined trends in Chinese older adults' cognitive function (Hu et al., 2021) and disability (Hu et al., 2022; Liang et al., 2017; Pan et al., 2023; Yu et al., 2016). The findings pointed out that ADL disability increased with age. A V-shaped cohort trend and a fluctuated period trend were identified by studying the Chinese samples. Females and urban residents were associated with more ADL limitations. When age increased, the gender and residence gaps in disability further increased. The cohort-based gender and residence inequalities in ADL limitations converged with successive birth cohorts. The age effect on trends in ADL was stronger as compared to period and cohort effects. Moreover, the gender and residence disparities in disability were found to rise with age and decreased with successive cohorts.

Regarding health dimensions of Chinese oldest old, particularly SRH, chronic diseases, cognitive function and subjective wellbeing, existing studies have not disentangled the period and cohort effects. So it is impossible determine why Chinese oldest old's health has changed during the past decades. Birth cohort membership may contribute the risk for younger birth cohorts over and above period effects that influence the society at large. Medical improvements and exposure to new technologies may prolong the life spans of individuals in new cohorts, and those individuals may have better health outcomes in general. Social changes and major social events may also have imprints on the oldest old's health outcome trajectories. Thus, age, period and cohort effects should exist when considering Chinese oldest old's health. To test the above research hypotheses, this study plans to utilize the APC approach and construct cross-classified random-effects models (CCREMS) to test the period and cohort effects when using demographic, socioeconomic and behavioral factors to predict health of Chinese oldest old.

3.2.2.3 The Age-Period-Cohort (APC) Approach and the Cross-Classified Random-Effects Models (CCREMs)

The conventional statistical APC analysis developed by Mason et al. (1973) focused on modeling age-by-time period tables of aggregate population data (for instance vital rates). Such a model, as Yang and Land (2008) argued, faces a major challenge of the underidentification problem induced by the exact linear dependency between age, period and cohort (Period = Age + Cohort) when the time intervals used to tabulate the data are of the same length for the age and period components. Yang and Land (2006) therefore suggested solving the underidentification problem in a repeated cross-section survey by using different temporal groupings for the age, period and cohort components. Usually, single years of age, time periods corresponding to years when surveys are conducted and cohorts that are defined by 5-year intervals are used to break the liner dependency at the individual respondent level. As a result, the fixed effect models in the APC approach are developed, which can estimate the age, period and cohort effects "by adding variables to control for the age, period and cohort effects in a conventional multiple linear regression analysis" (Yang, 2008a, b: 211).

Including more covariates in the individual level is also allowed in this case. The solutions for the underidentification problem along with the fixed effect models should take the multilevel structure of the data design into consideration. For example, in the CLHLS data that I merged and reconstructed by using eight waves of the survey data, an individual respondent is nested in and cross-classified by two higher level contexts-period and cohort. Raudenbush and Bryk (2002) have suggested that hierarchical linear models (HLM) can be used to analyze the cohort effect in age trajectories.

Considering the underidentification problem associated with conventional APC approach and the multilevel structure of the CLHLS data design, this research uses cross-classified random-effects models (CCREMs) to conduct empirical analyses. Individuals are the level-1 unit of analysis. Periods and cohorts operate as rows and columns in a matrix generating cohort-by-period cells, which become the level-2 unit of analysis. The CCREMs adjust for the dependency between age, period and cohort by considering period and cohort as cross-classified level-2 units. Beyond providing random cohort and period effects, CCREMs allow for random effects of independent variables, which is used to test for cross-cohort and cross-period variation in the effects of key independent variables (such as demographic and socioeconomic factors, health lifestyles et al.). Cohort- and period-specific random-effects coefficients suggest potential across-cohort and across-period changes in the effects of a variety of social and behavioral factors on Chinese oldest old's health. Fienberg and Mason (1982) have also proposed using a nonlinear transformation approach which applies nonlinear transformation such as polynomials to ensure the relationship of at least one of the age, period and cohort covariates to others is nonlinear. Following this strategy and noting that a finding of nonlinear age effect on health and well-being (Chen et al., 2010; Yang, 2008a, b), this study specifies models of health condition as a quadratic function of age. Specifically, the model estimates fixed effects of individual-level covariates and random effects of period (or time) and birth cohort. Individuals are the level-1 unit of analysis. Periods (or time) and birth cohort become the level-2 units of analysis. In my analyses of the CLHLS data, an individual respondent is nested in and cross-classified by a higher level context, which is period (or time) and birth cohort.

Beyond providing random period and cohort effects, CCREMs allow for random effects of independent variables, which is used to test cross-period or cross-cohort variations in the effects of key independent variables, i.e. a variety of social and behavioral variables. Period-specific random-effect coefficients suggest potential across-period or across-cohort changes in the effects of demographic, socioeconomic and behavioral variables on Chinese oldest old's health status.

Statistical models are presented in the analyses in each chapter. The study starts with the model that only includes age and age-squared as fixed effects. This model measures the overall effects of age, period and cohort on the oldest old's health. The following models further include demographic, socioeconomic and behavioral measures. Detailed models and the associated formula are presented in the following chapters. The analyses are performed by using SAS 9.4. The discussion of the APC approach and CCREMS allow researchers to replicate results presented in this manuscript.

3.3 Summary

This chapter introduces several available data sources that allow researchers to study health outcomes of Chinese older adults including the oldest old. The CLHLS dataset is the one to be examined in this manuscript. The chapter also elaborates the age-period-cohort (APC) approach and cross-classified random-effects models (CCREMs) that are used as the prime statistical methods in this research. The chapter also explains that since an individual respondent in the CLHLS surveys is nested in and cross-classified by two higher level contexts-period and cohort, the CCREMs are suitable to detect period and cohort effects when using a variety of demographic, socioeconomic and behavioral factors to predict health of Chinese oldest old. When constructing CCREMs, Chapters 4 through 9 in the manuscript employ a consistent format, which facilitates concordance between chapters and eases interpretation throughout the study. Chapter 4 and other following individual chapters will further detail the methodology and specific demographic, socioeconomic and health lifestyle measures used in the research.

References

Bardo, A. R., & Reither, E. N. (2021). Age-period-cohort models. In D. Gu & D. M. Dupre (Eds.), *Encyclopedia of gerontology and population aging* (pp. 188–194). Springer Publisher.

Barker, D. J. P. (1998). *Mothers, babies and health in later life*. Churchill Livingstone.

Chen, F., Yang, Y., & Liu, G. (2010). Social change and socioeconomic disparities in health over the life course in China: A cohort analysis. *American Sociological Review, 75*(1), 126–150.

Chen, X., Wang, Y., Strauss, J., & Zhao, Y. (2021). China health and retirement Longitudinal Study (CHARLS). In D. Gu & D. M. Dupre (Eds.), *Encyclopedia o Gerontology and Population Aging*. Springer Publisher.

Elder, G. H. (1987). War mobilization and the life course: A Cohort of World War II Veterans. *Sociological Forum, 2*(3), 449–472.

Fienberg, S. E., & Mason, W. M. (1982). *Specification and Implementation of Age, Period and Cohort Models*. Springer-Verlag.

Firebaugh, G. (1989). Methods for estimating cohort replacement effects. *Sociological Methodology, 19*, 243–262.

Fogel, R. W., & Costa, D. L. (1997). A theory of technophysio evolution, with some implications for forecasting population, health care costs, and pension costs. *Demography, 34*(1), 49–66.

Glenn, N. D. (2003). Distinguishing age, period, and cohort effects. In J. T. Mortimer & M. J. Shanahan (Eds.), *Handbook of the life course* (Vol. VI, A, pp. 465–476). Springer Publisher.

Guo, P., & Li, K. (2012). Trends in esophageal cancer mortality in China during 1987–2009: Age, period and birth cohort analyzes. *Cancer Epidemiology, 12*, 003.

Han, J. H., & Zhao, X. Q. (2021). Impact of internet use on multi-dimensional health: an empirical study based on CGSS 2017 data. *Frontiers in Public Health, 749*, 816.

Hu, X., Gu, S., Zhen, X., Sun, X., Gu, Y., & Dong, H. (2021). Trends in cognitive function among Chinese elderly From 1998 to 2018: An age-period-cohort analysis. Frontiers in Public Health, 753871.

Hu, X., Gu, S., Zhen, X., Sun, X., Gu, Y., & Dong, H. (2021). Trends in cognitive function among Chinese elderly From 1998 to 2018: An age-period-cohort analysis. *Frontiers in Public Health, 753871*.

Hu, X., Gu, S., Zhen, X., Sun, X., Gu, Y., & Dong, H. (2022). Trends in activities of daily living disability among Chinese older adults from 1998 to 2018: An age-period-cohort analysis. *European Journal of Ageing, 19*(4), 1167–1179.

Li, T., & Zhang, Y. (2023). Intergenerational relationships across multiple children and older parents' depressive symptoms in China—A resource contingency perspective. *Journals of Gerontology: Social Sciences*. Published online.

Li, M., Wan, X., & Wang, Y. (2017). Time trends of esophageal and gastric cancer mortality in China, 1991–2009: An age-period-cohort analysis. *Scientific Reports, 7*(1), 6797.

Liang, Y., Welmer, A., Möller, J., & Qiu, C. (2017). Trends in disability of instrumental activities of daily living among older Chinese adults, 1997–2006: population based study. *BMJ Open, 7*(8), e016996.

Liu, X., Yu, Y., Wang, M., & Mubarik, S. (2020). The mortality of lung cancer attributable to smoking among adults in China and the United States during 1990–2017. *Cancer Communications, 40*(11), 611–619.

Mason, K. O., Mason, W. M., Winsborough, H. H., & Poole, W. K. (1973). Some Methodological Issues in Cohort Analysis of Archival Data. *American Sociological Review, 38*(2), 242–258.

Pan, C., Cao, N., Kelifa, M., & Luo, S. (2023). Age and cohort trends in disability among Chinese older adults. *Frontiers in Public Health, 11*(998), 948.

Raudenbush, S. W., & Bryk, A. S. (2002). *Hierarchical linear models: Applications and data analysis methods* (2nd ed.). Sage Publications.

Reither, E. N., Hauser, R. M., & Yang, Y. (2009). Do birth cohorts matter? Age-period-cohort analyses of the obesity epidemic in the United States. *Social Science & Medcine, 69*(10). https://doi.org/10.1016/j.socscimed.2009.08.040

Ryder, N. B. (1965). The cohort as a concept in the study of social change. *American Sociological Review, 30*(6), 843–861.

Schwadel, P. (2010). Age, period, and cohort effects on U.S. Religious service attendance: the declining impact of sex, southern residence, and catholic affiliation. *Sociology of Religion, 71*(1), 2–24.

Sun, Y., Wang, Y., & Li, M. (2018). Long-term trends of liver cancer mortality by gender in urban and rural areas in China: An age-period-cohort analysis. *BMJ Open, 8*(2), e020490.

Yang, Y. (2008a). Age, period, cohort effects. In D. Carr (Ed.), *Encyclopedia of the life course and human development* (Vol. 3, pp. 6–10). Macmillan Reference USA.

Yang, Y. (2008b). Social inequalities in happiness in the United States, 1972 to 2004: An Age-Period-Cohort Analysis. *American Sociological Review, 73*(2), 185–356.

Yang, Y., & Land, K. C. (2006). A mixed models approach to the age-period-cohort analysis of repeated cross-section surveys, with an application to data on trends in verbal test scores. *Sociological Methodology, 36*(1), 75–97.

Yang, Y., & Land, K. C. (2008). Age–period–cohort analysis of repeated cross-section surveys: Fixed or random effects? *Sociological Methods & Research, 36*(3), 297–326.

Yang, S. Y., Xie, L. L., & Li, T. (2021). Living arrangements and intergenerational support among older Chinese: Does the gender of the co-resident adult children make a difference? *International Social Work*, Published online.

Yu, R., Wong, M., Chang, B., Lai, X., Lum, C., Auyeung, T., Lee, J., Tsoi, K., Lee, R., & Woo, J. (2016). Trends in activities of daily living disability in a large sample of community-dwelling Chinese older adults in Hong Kong: an age-period-cohort analysis. *BMJ Open, 6*(12), e013259.

Zeng, Y., Vaupel, J., Xiao, Z., Zhang, C., & Liu, Y. (2002). Sociodemographic and health profiles of the oldest old in China. *Population and Development Review, 28*, 251–273.

Zhang, B., Zhai, F. Y., Du, S. F., & Popkin, B. M. (2014). The China Health and Nutrition Survey,1989–2011. *Obesity Reviews*, *Supp*(1), 2–7.

Zhang, B., Wang, H., & Du, S. (2021). China Nutrition and Health Surveys (CHNS). In D. Gu & D. M. Dupre (Eds.), *Encyclopedia of gerontology and population aging* (pp. 943–948). Springer Publisher.

References

Part II

Chinese Oldest Old's Health Status and Its Determinants

Changing Patterns of the Oldest Old's Health by Period and across Birth Cohort

4

Starting from this chapter, the manuscript moves to empirical analyses of Chinese oldest old's health and its determinants through analyzing data derived from eight waves of CLHLS. The manuscript pays a particular attention to period and cohort variations in the links between various demographic, socioeconomic and behavioral covariates and the oldest old's health. Before moving to Chaps. 5, 6, 7, 8, and 9, this chapter first describes how Chinese oldest old's health status has changed during 1998 to 2018 by period and across birth cohorts. The chapter then uses CCREMs to show period and cohort variations when only including age, age-squared term, period and cohort in regression models to predict Chinese oldest old's health.

4.1 Data, Health Measures and Methods

4.1.1 Data

This chapter examines data from eight (1998, 2000, 2002, 2005, 2008, 2011–12, 2014 and 2018) waves of the Chinese Longitudinal Healthy Longevity Survey (CLHLS). The CLHLS has been considered as the first large survey of the oldest old conducted in a developing country (Zeng et al., 2002). The data were collected by Peking University's Center for Healthy Aging and Family Studies and the China National Research Center on Aging, with support from the U.S. National Institute on Aging. The CLHLS was undertaken in 631 randomly selected counties and cities of 22 provinces and municipalities directly under the Central Government in China. These 22 provinces and municipalities include Liaoning, Jilin, Heilongjiang, Hebei, Beijing, Tianjin, Shanxi, Shaanxi, Shanghai, Jiangsu, Zhejiang, Anhui, Fujian, Jiangxi, Shandong, Henan, Hubei, Hunan, Guangdong, Guangxi, Sichuan, and Chongqing. The survey covers roughly half the counties and cities of those provinces, and the sample areas represent 85.0% of the total population of China.

L. Zhang, *Living Longer and Healthier at Older Ages*, Advances in Studies of Aging and Health 3, https://doi.org/10.1007/978-3-031-69773-9_4

The1998 baseline survey includes 9,073 oldest old individuals aged 80 and older. In 2000, 4,844 of those elders were reinterviewed, and 6,372 elders were newly added interviewees. For more detailed description of the data, see Zeng et al. (2002). This manuscript studies the oldest old aged 85 to 105 since prior literature considered persons who reported age 106 or higher as invalid cases (Zeng et al., 2002).

4.1.2 Health Measures

Health outcomes in this study are measured by five indictors, which are consistent with measures used in previous research, including self-rated health (SRH), number of chronic diseases, cognitive function, and positive and negative feelings which are considered as measures of one's subjective wellbeing (Lv et al., 2019; Oftedal et al., 2019; Smith & Victor, 2019; Yuan, 2021). The respondent's *self-rated health* is coded as a continuous variable (1 = very bad, 5 = very good). *Number of chronic diseases* is measured by the number of chronic diseases reported by the respondent when the survey was conducted, which ranges from 0 to 13. *Cognitive function* of the respondent is measured by using the Chinese version of the Mini-Mental State Examination (MMSE). The MMSE is adapted from Folstein, Folstein, and McHugh (Folstein et al., 1975) and tests four aspects of cognitive functioning: orientation, calculation, recall, and language. The total possible score on the MMSE is 30, with lower scores indicating poor cognitive ability. Responses of "unable to answer" are coded as incorrect answers.

Regarding subjective well-being, the study relies on a number of CLHLS questions which measured positive and negative feelings of the respondent, respectively. Since the study conducts longitudinal analyses, questions used in the 1998 CLHLS survey are considered as baseline questions. Although a number of additional questions have been added in later years, the study only relies on those baseline questions to perform the analyses. Thus, the questions asking positive feelings include: (1) Do you always look on the bright side of things? (2) How do you rate your life at present? (3) Are you happy now as when you were younger?

The questions asking negative feelings are: (1) Do you often feel fearful or anxious? (2) Do you often feel lonely and isolated? (3) Do you feel the older you get the more useless you are? The responses range from 1 to 5 for the questions asking positive as well as negative feelings with "1" representing the lowest level and "5" representing the highest level. Since CLHLS data were not collected to examine the psychological well-being of the elderly, the above questions may not be perfect indicators of one's subjective well-being. However, Chen and Short (Chen & Short, 2008, p. 1388) indicated that: "they represent important dimensions of subjective well-being, such as life satisfaction, happiness, and loneness." Thus, measures associated with the above questions are considered as legitimate indicators of the oldest old psychological well-being.

In this analysis, two indices are created by adding questions asking positive and negative feelings of the respondents. The study adds answers to questions asking positive feelings 1 to 3 as an index representing positive well-being, adding items asking negative feelings 1 to 3 as an index of negative well-being. Thus, the indices are summary scores for each set of variables discussed above, which is created by adding the raw scores of all variables in each set. Accordingly, the score for each index ranges from 3 to 15. The logic behind the strategy is that each group of variables measures the same concept. This strategy reduces the number of variables in the analysis and improved the efficiency of the regression models. After summing each set of variables to a single variable, Cronbach's alpha is applied to assess the reliability of a given set of variables (Cronbach, 1951).

4.1.3 Methods

This analysis considers multiple waves of CLHLS data as repeated cross-section surveys. A cross-classified random-effects model (CCREM) is used to take into account the embeddedness of respondents in the CLHLS surveys within a time period (Yang & Land, 2006). Specifically, the model estimates fixed effects of individual-level covariates and random effects of period (or time) and birth cohort. Individuals are the level-1 unit of analysis. Periods (or time) and birth cohort become the level-2 units of analysis. In the analysis of the CLHLS data, an individual respondent is nested in and cross-classified by a higher level context, which are period (or time) and birth cohort.

Beyond providing random period effects, CCREMs allow for random effects of independent variables, which is used to test for cross-period variations in the effects of key independent variables. Period-specific random-effects coefficients and cohort-specific random-effects coefficients suggest potential across-period and across-cohort changes in the effects of independent variables on the oldest old's health status, respectively. This study specifies models of the oldest old's health status as a quadratic function of age.

In this chapter, only models including age variables, period and cohorts are presented in the analysis. The first model only includes age and age-squared as fixed effects. This model measures the overall effects of period and cohort on the oldest old's health status, controlling for age and age squared variables. The full individual level or level-1 equation can be expressed as:

$$Y_{ijk} = \beta_{0jk} + \beta_{1jk}AGE + \beta_{2jk}AGE^2 + \beta_{3jk}X + \sum_{p=10}^{p} \beta_p Xp + e_{ijk} \tag{4.1}$$

where Y_{ijk} represents health status of the *i*th respondent for i = 1, …, n*j* individuals within the *j*th period for j = 1, …. β_{0j} is the intercept or for respondents in period *j*; "AGE" and "AGE^2" symbolize age and age-squared, respectively. "X" indicates independent variables. β_p represents other individual-level fixed effects where *k* is the maximum number of covariates. e_{ij} represents an individual level random error term.

The level-2 model can be expressed as:

$$\beta_{0j} = \gamma_0 + p_{0j} + c_{0k} \tag{4.2}$$

This model specifies the overall mean vary from period to period and from cohort to cohort. γ_0 is the model intercept, which is the expected mean at the zero value of all level-1 variables averaged over all periods and cohorts. Here p_{0j} and c_{0k} are the residual random effects of period and cohort, respectively. The cell mean, $\beta_{0j,}$ is equal to the sum of the overall mean or intercept (γ_0), and the residual random effect of period j (p_{0j}) and the residual random effect of cohort (c_{0k}). The residual random effects of period j (p_{0j}) and cohort k (c_{0k}) examine the effects of each period and cohort on a specific health measure, respectively. When adding (4.2) to (4.1), it yields the level-1 and level-2 combined model:

$$Y_{ijk} = \gamma_0 + \beta_{0jk} + \beta_{1jk}AGE + \beta_{2jk}AGE^2 + \beta_{3jk}X + p_{0j} + c_{0k} + e_{ijk} \qquad (4.3)$$

where, within each survey year j, respondent i's health is modeled as a function of his or her age, age-squared, the independent variable and other control variables. In the above model, γ_0 represents the fixed-effects coefficients for independent variables. The symbols of p_{0j} and c_{0k} represent the period-specific and cohort-specific effects of independent variables, respectively. The model tests for random effects of certain independent variables on the respondent's health across periods and cohorts. All continuous variables are sorted and centered around their overall mean.

Since there are eight waves of survey data, survey years for the eight waves are considered as the eight periods. I have developed 13 birth cohorts in the analyses (pre-1897, 1898–1900, 1901–03, 1904–06, 1907–09, 1910–12, 1913–15, 1916–18, 1919–21, 1922–24, 1925–27, 1928–30 and 1931–33). The 1934 and later birth cohorts are dropped from the analyses since those cohorts do not fall into the age range of 85 to 105. When showing descriptive results, all 13 birth cohorts are shown in graphs. When conducting regression analyses, the 1928–30 and 1931–33 birth cohorts are combined into one birth cohort considering the number of respondents who were born in and after 1928 is limited.

4.2 Descriptive Results

4.2.1 Basic Statistics of Variables

Table 4.1 presents descriptive information for all variables used in this research, including those used in other chapters of this manuscript. Since CLHLS is a sampling survey and the study does not use the sampling weight, the percentage distributions shown in the study refer to the sample. In total, 58,485 oldest old have been surveyed in eight waves of CLHLS. In terms of the demographic profiles, the mean age of the oldest old is 94.3 with 47.0% of respondents being urban and 53.0% of them being rural. About 13.9% of the oldest old reported themselves as being married. The respondents' SES is low with the average schooling year being 1.4. The majority of the respondents reported elementary school education or less. And 67.9% of them reported that they frequently went to bed hungry in childhood.

Table 4.1 Descriptive statistics of variables: Chinese aged 85–105 (%) (Unweighted)

Variables	Total	N
Dependent variables		
Physical health		
Self-rated health		
Very bad	1.48	50,847
Bad	13.31	
Fair	35.73	
Good	38.65	
Very good	10.83	
Number of chronic diseases (mean)	1.03	55,114
Mental health		
Cognitive function score (mean)	21.63	49,168
Positive feeling score (mean)	10.38	50,662
Negative feeling score (mean)	7.02	48,249
Demographic and SES variables		
Age (mean)	94.33	58,485
Sex		
Male	38.03	58,485
Female	61.97	
Residence		
Urban	47.03	58,485
Rural	52.97	
Schooling years (mean)	1.37	56,665
Marital status		
Currently married	13.93	58,191
Currently not married	86.07	
Often going to bed hungry as a child		
Yes	67.90	56,183
No	32.10	
Other main independent variables		
Healthy lifestyles		
Smoking in past and still a smoker		
Yes	11.98	58,289
No	88.02	
Drinking in past and still a drinker		
Yes	13.69	58,199
No	86.31	
Exercising in past and at present		
Yes	16.50	58,156
No	83.50	
Having leisure activities everyday		
Yes	38.45	58,217
No	61.55	

Sources: 1998–2018 Chinese Longitudinal Healthy Longevity Survey (CLHLS)

Regarding health outcomes of the oldest old, Table 4.1 indicates that the majority of the respondents rated their overall health as "fair" (35.7%) or "good" (38.7%). The self-reported average number of chronic diseases is 1.0. The mean cognitive function score is 21.6. Since the cognitive function scores range from 0 to 30, the mean cognitive function score of 21.6 suggests a fairly good cognitive function of the studied oldest old. The mean positive and negative feeling scores are 10.4 and 7.0, respectively. Considering the possible scores for positive and negative feelings are from 3 to 15, the mean positive and negative feeling scores imply a fairly good subjective wellbeing status of the studied oldest old. Detailed descriptive information for period and cohort variations in these variables will be depicted in graphs in the following sections. More detailed descriptive statistics for main independent variables will be presented in following chapters.

4.2.2 Period and Cohort Variations in Health Outcomes

The changing patterns of Chinese oldest old's health outcomes are demonstrated in Figs. 4.1, 4.2, 4.3, 4.4, and 4.5. Since SRH is a coded as a continuous variable with values ranging from 1 to 5. The study focuses on respondents who rated their health as "very good" to perform descriptive analyses of the SRH variable. In addition to presenting figures, Table 4.2 is also constructed to show the multilevel data structure of the studied sample. The table demonstrates the percentage distribution of those respondents who rated their health as "very good." As it is shown here, respondents with"very good" SRH status are nested in, and cross-classified by the two

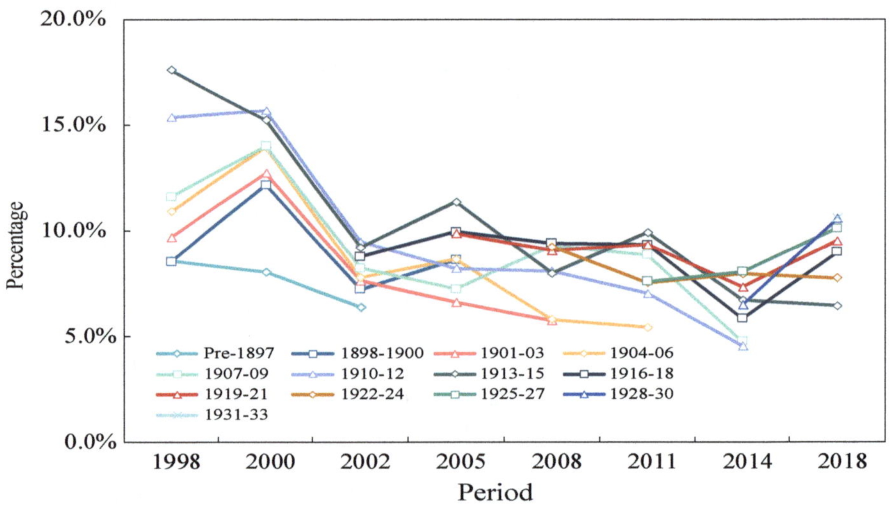

Fig. 4.1 Percentage of respondents who rated their health as "very good" by time period and across birth cohorts (Unweighted). (Sources: 1998–2018 Chinese Longitudinal Healthy Longevity Survey (CLHLS))

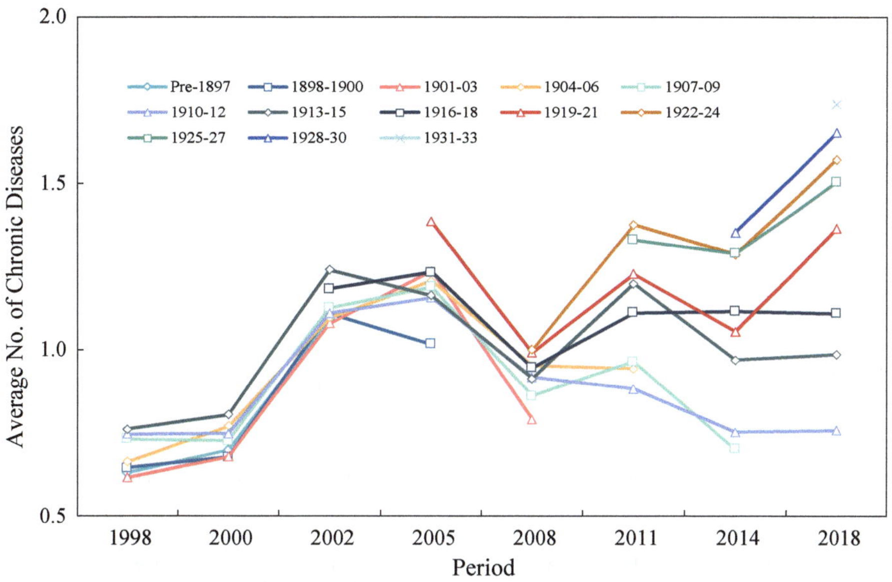

Fig. 4.2 Average number of reported chronic diseases by Time Period and across birth cohorts (Unweighted). (Sources: 1998–2018 Chinese Longitudinal Healthy Longevity Survey (CLHLS))

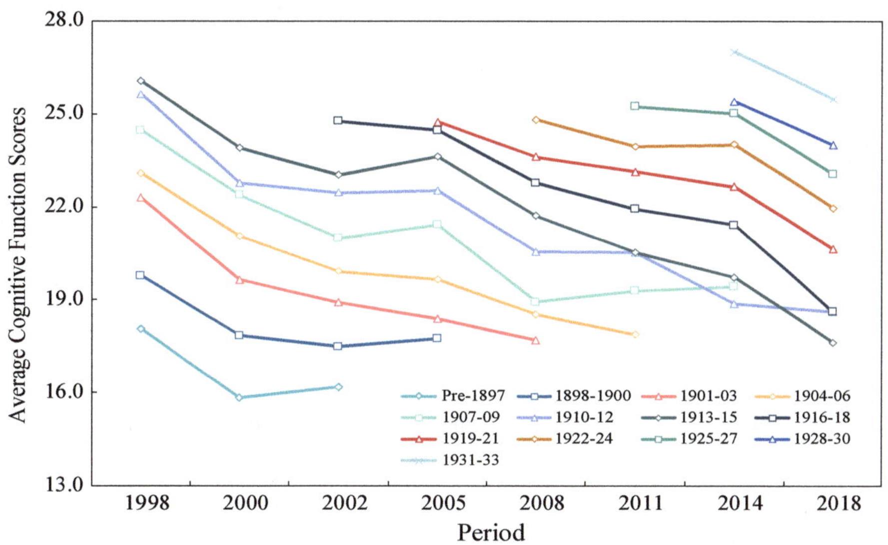

Fig. 4.3 Average cognitive function score of respondents by time period and across birth cohorts (Unweighted). (Sources: 1998–2018 Chinese Longitudinal Healthy Longevity Survey (CLHLS))

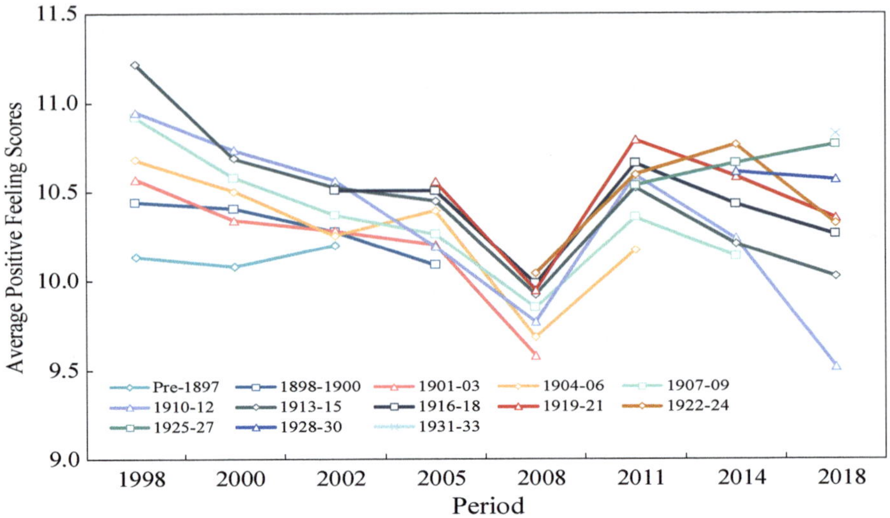

Fig. 4.4 Average positive feeling score of respondents by time period and across birth cohorts (Unweighted). (Sources: 1998–2018 Chinese Longitudinal Healthy Longevity Survey (CLHLS))

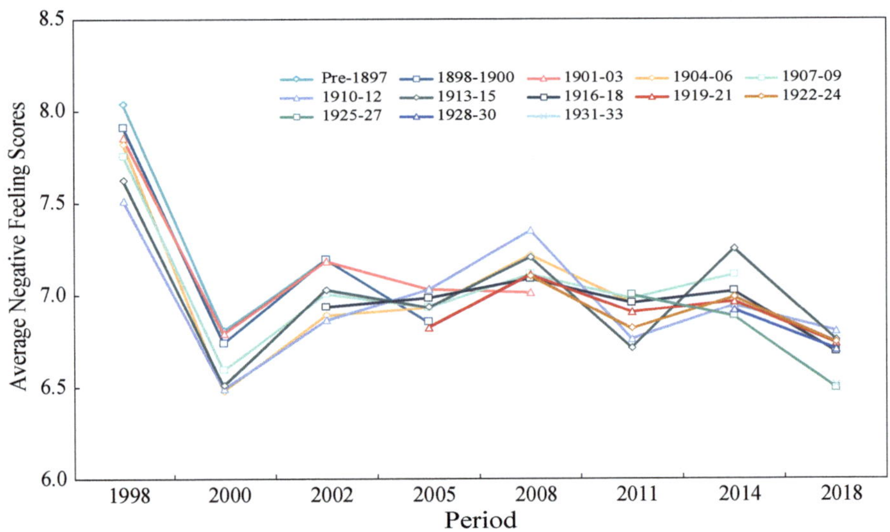

Fig. 4.5 Average negative feeling score of respondents by time period and across birth cohorts (Unweighted). (Sources: 1998–2018 Chinese Longitudinal Healthy Longevity Survey (CLHLS))

higher-level social contexts defined by time period and birth cohort. Respondents of any birth cohort can be interviewed in multiple surveys, and respondents in any particular wave of the survey can also be drawn from multiple birth cohorts.

The variations in respondents' SRH is displayed in Fig. 4.1. As the results show, overall, less than one-fifth of the respondents rated their health as"very good".

Table 4.2 Percentages of respondents who self-rated their health as "Very Good" by period and birth cohort: Chinese aged 85–105 (Unweighted)

	Period									
	1998–2000		2002–2005		2008–2011		2014–2018		Total	
Cohort	%	N	%	N	%	N	%	N	%	N
Pre-1897	8.42	233	6.38	18					8.23	251
1898–1900	10.70	277	7.49	134					9.39	411
1901–03	11.19	181	7.21	201	5.74	14			8.52	396
1904–06	12.51	251	8.33	189	5.77	88			9.10	528
1907–09	13.06	328	7.83	183	9.19	204	4.76	3	10.07	718
1910–12	15.55	355	8.86	235	7.71	115	4.01	12	10.66	717
1913–15	15.44	216	10.44	295	8.52	181	6.54	68	10.29	760
1916–18			9.59	206	9.38	271	8.33	178	9.13	655
1919–21			9.88	79	9.15	197	8.52	95	9.12	371
1922–24					8.38	128	7.82	123	8.10	251
1925–27					7.60	37	9.38	167	8.99	204
1928–33					1.57	3	9.42	241	8.88	244
Total	12.13	1,841	8.61	1,540	8.34	1,238	8.49	886	10.83	5,506

Sources: 1998–2018 Chinese Longitudinal Healthy Longevity Survey (CLHLS)

Clearly, lines representing various birth cohorts intertwine with each other. Except for the 1928 to 1933 birth cohort, younger birth cohorts seem to be more inclined to self-rate their health as "very good" as compared to older birth cohorts. Regarding variations in SRH by time period, although some fluctuations occur, a general declining trend is shown during 1998 to 2014. That is, before year 2014, for a specific birth cohort, as time goes by, the percentage of the oldest old who reported their health as "very good" tends to decline. Since 2014, percentages of respondents who rated their health as "very good" turn to increase for several birth cohorts.

Figure 4.2 depicts changing patterns of the average number of chronic diseases among Chinese oldest old. The average numbers of choric diseases reported by the oldest old range from 0.6 in 1998 to 1.7 in 2018 with an increasing trend being observed across birth cohorts. The differentials become more substantial since year 2008 with more recent birth cohorts reporting greater average numbers of chronic diseases as compared to earlier birth cohorts. As to variations in chronic diseases by time period, it appears that respondents in years 2002 to 2005 reported greater average numbers of chronic diseases for various birth cohorts. The average numbers of chronic diseases then decline in years 2006 to 2008 and rise again since year 2008. The rising trend of chronic diseases may be explained by China's transition from infectious diseases to chronic diseases.

The average cognitive function scores by time period and across birth cohorts are presented in Fig. 4.3. The average cognitive function scores range between 16.2 to 27.0. Since the highest value possible for cognitive function score is 30, Chinese oldest old demonstrate fairly good cognitive function. In addition, younger birth cohorts show a better cognitive function status than earlier birth cohorts. It is interesting to see that except for the 1928 to 1938 birth cohort, all other birth cohorts show declining trends of respondents' cognitive function over time. It means that as time goes by, the oldest old's cognitive function score declines in more recent years. It makes sense because the studied respondents are aging over time. Accordingly, the figure shows declining trends in respondents' cognitive function.

The average positive and negative feeling scores are presented in Figs. 4.4 and 4.5, respectively. As it was pointed out in previous section, the possible positive and negative feeling scores range from 3 to 15. According to Fig. 4.4, the average positive feeling scores for respondents of various birth cohorts range from 9.5 to 11.2, indicating a fairly positive subjective wellbeing status of the oldest old. Lines representing mean positive feeling scores for different birth cohorts intertwine with each other.

No significant birth cohort differentials are shown here; more recent birth cohorts only report slightly higher average positive feeling scores than earlier birth cohorts. It looks like the variations in positive feeling scores among different birth cohorts are more apparent before and after year 2008. The variations among birth cohorts shrink in year 2008 and reach the lowest levels among all birth cohorts. It indicates that year 2008 is a unique year that makes birth cohort differentials almost disappear.

As far as variations in time period, it shows that before year 2008, the average positive feeling scores have been declining over time. After 2008, the mean positive feeling scores for various birth cohorts increase and reach their peak in year 2011 for most birth cohorts. The lowest positive feeling scores in year 2008 may relate to the *Wenchuan* earthquake and the economic depression that China experienced during that time. It suggests that the oldest old's subjective welling can be heavily impacted by an economic hardship and natural disasters of a country.

The average negative feeling scores presented in Fig. 4.5 are from 6.5 to 8.0 for the studied birth cohorts. Since the possible negative feeling scores range from 3 to 15, the average negative feeling scores reported by the respondents suggest a fairly good subjective wellbeing status of the studied Chinese oldest old. Lines representing different birth cohorts intertwine with each other without showing significant cohort differences. When the time effect is considered, years 2000 and 2018 are the ones that show the lowest average negative feeling scores; years 2002, 2008 and 2014 appear to show higher negative feeling scores for sampled birth cohorts. In short, period variations are more apparent than cohort variations in terms of Chinese oldest old's negative feeling scores. These findings imply that Chinese oldest old's subjective wellbeing is more likely to be affected by social events that are associated with time period changes than birth cohort variations.

Overall, the findings show that Chinese oldest old's SRH, negative feeling scores and cognitive function scores tend to decline over time. But the average number of chronic diseases increases over time. The positive feeling scores show a "V" shape, indicating significant period influence on Chinese oldest old's subjective wellbeing. Cohort wise, the average number of chronic diseases and cognitive function scores are in a rising trend across birth cohorts. Both period and cohort variations are apparent for other health measures.

4.3 Regression Results Based on APC Analyses

This part of the chapter switches to presenting regression results by performing APC analyses when only including period, cohort, age and age-squared term. Table 4.3 presents findings of estimated health outcomes, measured by SRH, cognitive function, subjective wellbeing and chronic diseases from the CCREMs. Each model presented in this table only includes period, cohort covariates along with age and age-squared term. The models indicate the overall impact of age, period and cohort on health measures. The fixed-effects coefficients presented in the models can be interpreted in a manner similar to ordinary least square (OLS) coefficients. Some key results are also illustrated by the predicted levels of health measures in graphs. All tables present estimates of fixed-effects coefficients and random-effects variance components from the CCREMs.

The chapter first interprets results that are presented in Table 4.3 which estimate when only age and age-squared terms are included in the CCREMs, the overall trends of health outcomes by period and across birth cohort. The age effect appears to be significant. But the results in Table 4.3 show a non-significant quadratic age effect on the oldest old's health net of the random period and cohort effects. It implies that the relationships between age and health measures are linear. That is, adjusting for time period and birth cohort variations, with age increasing, the level of SRH increases but the levels of other health dimension decline.

The estimates of random effects in terms of residual variance components at level-2 suggest significant period and cohort effects when controlling for the age effect, as reported in the lower panel of Table 4.3. Obviously, net of the age effects, the predicted SRH varies significantly by time period and across birth cohorts. Figures 4.6a, 4.6b, 4.7a, 4.7b, 4.8a, 4.8b, 4.9a, 4.9b, 4.10a, and 4.10b are generated based on regression results from the CCREMs in Table 4.3 to depict the overall trends of predicted health, measured by SRH, chronic diseases, cognitive function and subjective wellbeing.

Figure 4.6a displays the estimated period effects, namely, the predicted SRH for each year at the mean age over all birth cohorts. The period effects show a general weak downward trend across the first two periods, the period effects then increase slightly and fluctuate afterwards. The period effects exhibit clear nonlinear declines over time. Years 2002 and 2014 appear to be the years that show lower predicted SRH among all time periods. The magnitude of such period changes is relatively small since the predicted SRH largely fell between 3.4 and 3.6 during the survey years.

Figure 4.6b shows the estimated cohort effects in terms of the predicted SRH at the mean age over all periods. Similar to the period effects, the cohort effects are small. The cohort effects first show an increasing trend prior to the 1910 to 1912 cohort, and a declining trend is observed for later birth cohorts. Obviously, the cohort changes are smaller as compared to period variations.

Figures 4.7a and 4.7b show the period and cohort effects, respectively, when predicting the number of chronic diseases. Figure 4.7a shows the predicted numbers of chronic diseases change from 0.6 in 1998 to 1.4 in 2018, indicating an increasing

Table 4.3 Regression results of health outcome measures on age from CCREM models: Chinese aged 85–105

	Self-Rated Health		Chronic disease		Cognitive function		Positive feelings		Negative feelings	
Variables	Coef.	P Value	Coef.	P Value	Coef.	P Value	Coef.	P Value	Coef.	P Value
Fixed effect component										
Intercept	3.430	<0.001	1.071	<0.001	21.495	<0.001	10.417	<0.001	7.038	<0.001
Control variables										
Age	0.000	0.978	−0.019	<0.001	−0.418	<0.001	−0.028	<0.001	0.011	<0.001
Age^2	−0.000	0.340	−0.000	0.212	−0.001	0.442	−0.001	0.061	−0.000	0.382
Random effect component										
Period										
Intercept	0.009	0.034	0.071	0.033	0.414	0.038	0.050	0.033	0.142	0.032
Cohort										
Intercept	0.001	0.065	0.007	0.024	0.133	0.026	0.003	0.117	0.002	0.112
AIC	133,688.9		183,779.3		329,519.5		232,636.1		218,468.3	
BIC	133,682.9		183,773.3		329,513.5		232,630.1		218,462.3	

Sources: 1998–2018 Chinese Longitudinal Healthy Longevity Survey (CLHLS)

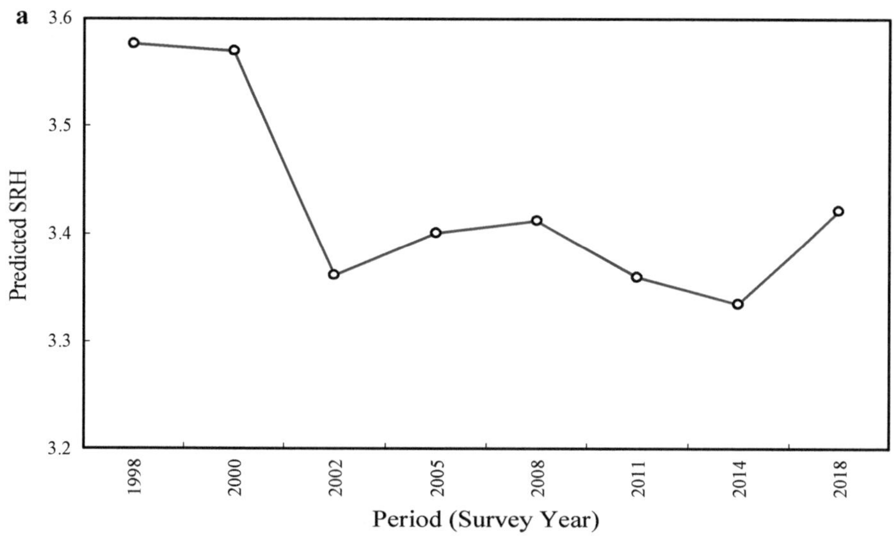

Fig. 4.6a Predicted SRH over time period. (Sources: 1998–2018 Chinese Longitudinal Healthy Longevity Survey (CLHLS))

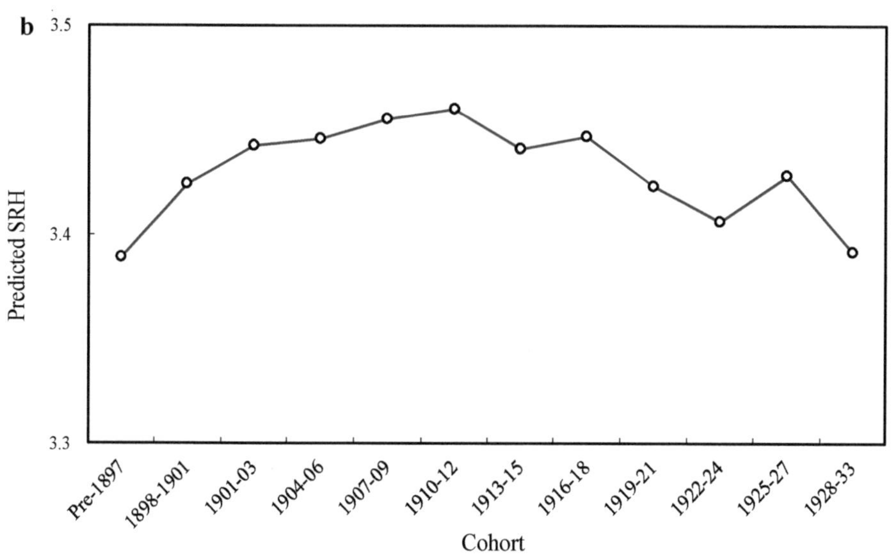

Fig. 4.6b Predicted SRH over birth cohort. (Sources: 1998–2018 Chinese Longitudinal Healthy Longevity Survey (CLHLS))

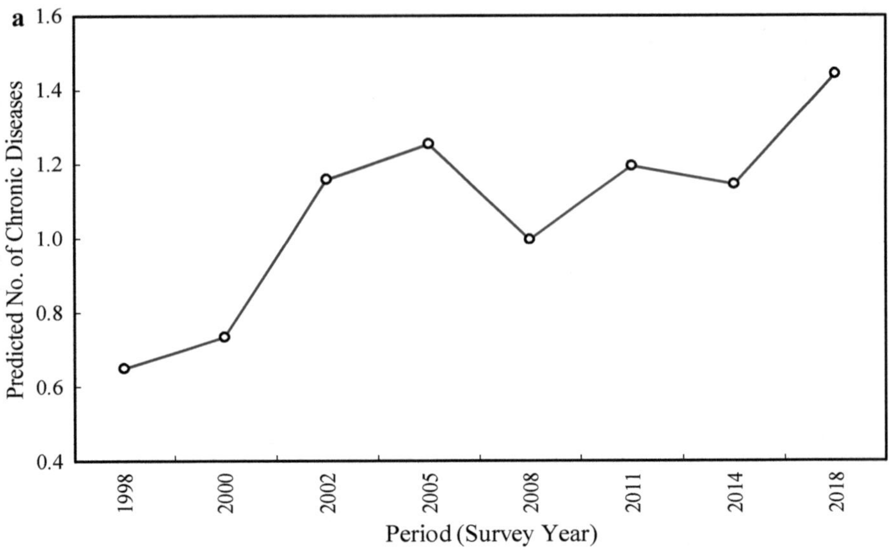

Fig. 4.7a Predicted number of chronic diseases over time period. (Sources: 1998–2018 Chinese Longitudinal Healthy Longevity Survey (CLHLS))

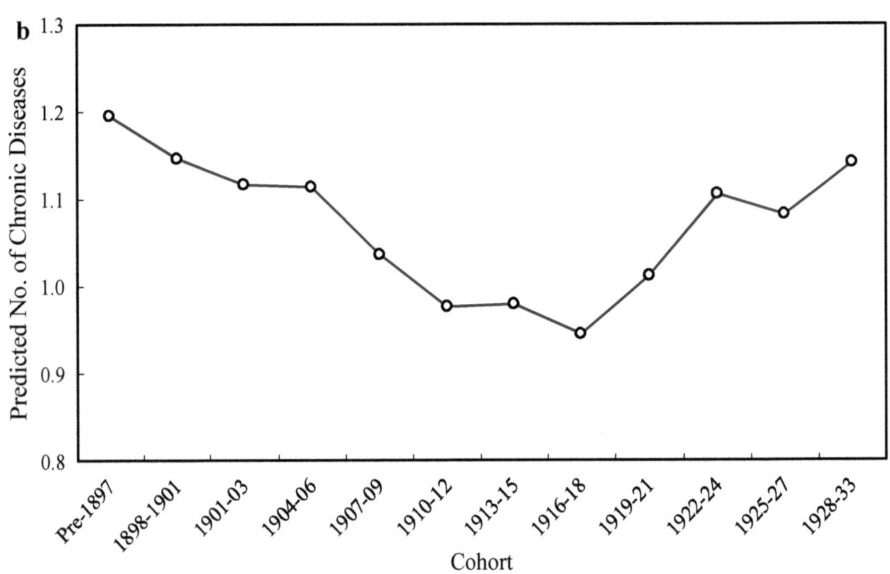

Fig. 4.7b Predicted number of chronic diseases over birth cohort. (Sources: 1998–2018 Chinese Longitudinal Healthy Longevity Survey (CLHLS))

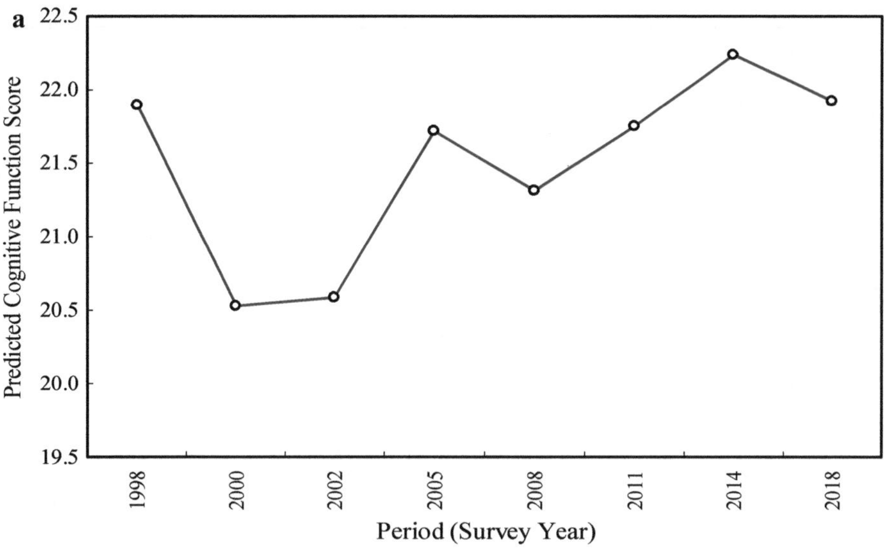

Fig. 4.8a Predicted cognitive function score over time period. (Sources: 1998–2018 Chinese Longitudinal Healthy Longevity Survey (CLHLS))

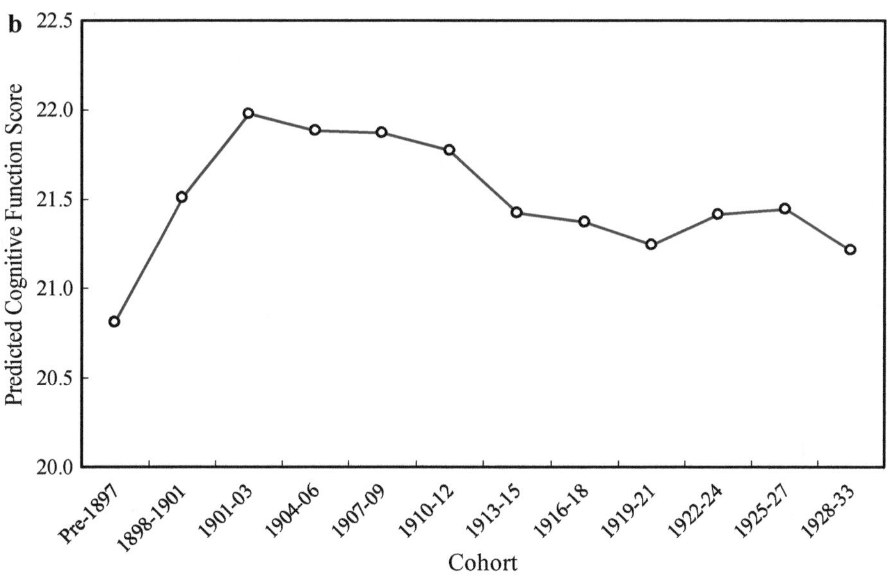

Fig. 4.8b Predicted cognitive function score over birth cohort. (Sources: 1998–2018 Chinese Longitudinal Healthy Longevity Survey (CLHLS))

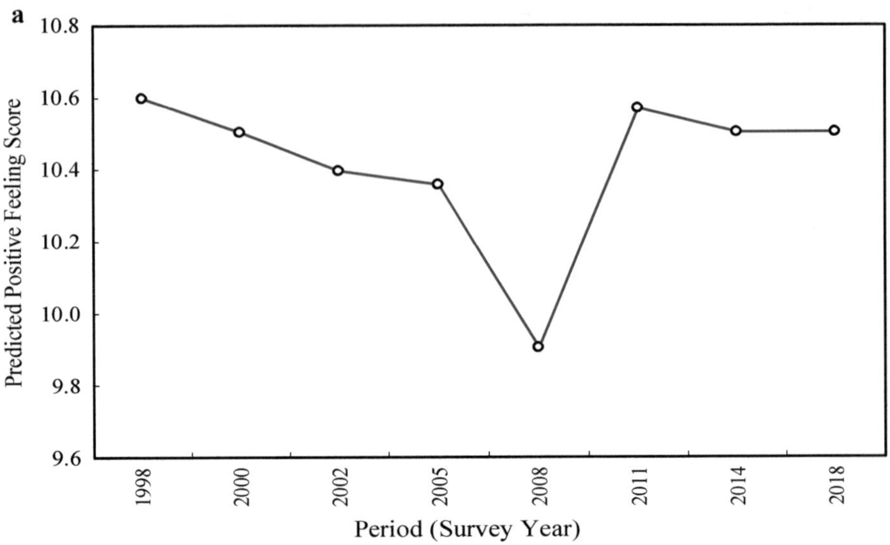

Fig. 4.9a Predicted positive feeling score over time period. (Sources: 1998–2018 Chinese Longitudinal Healthy Longevity Survey (CLHLS))

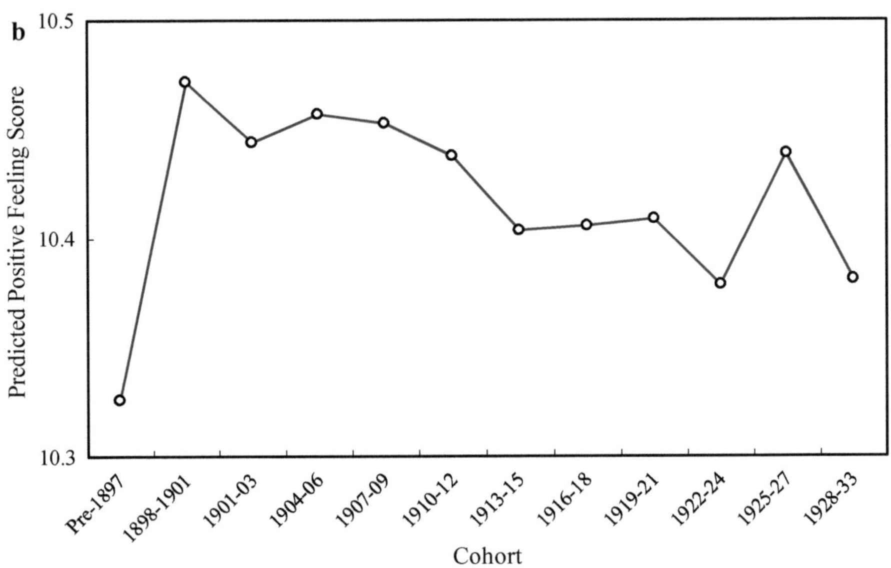

Fig. 4.9b Predicted positive feeling score over birth cohort. Sources: 1998–2018 Chinese Longitudinal Healthy Longevity Survey (CLHLS))

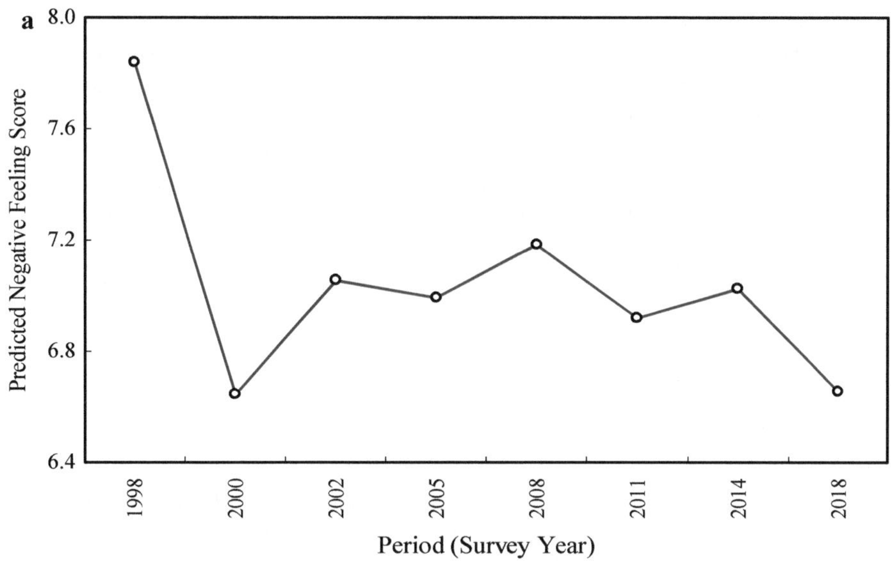

Fig. 4.10a Predicted negative feeling score over time period. (Sources: 1998–2018 Chinese Longitudinal Healthy Longevity Survey (CLHLS))

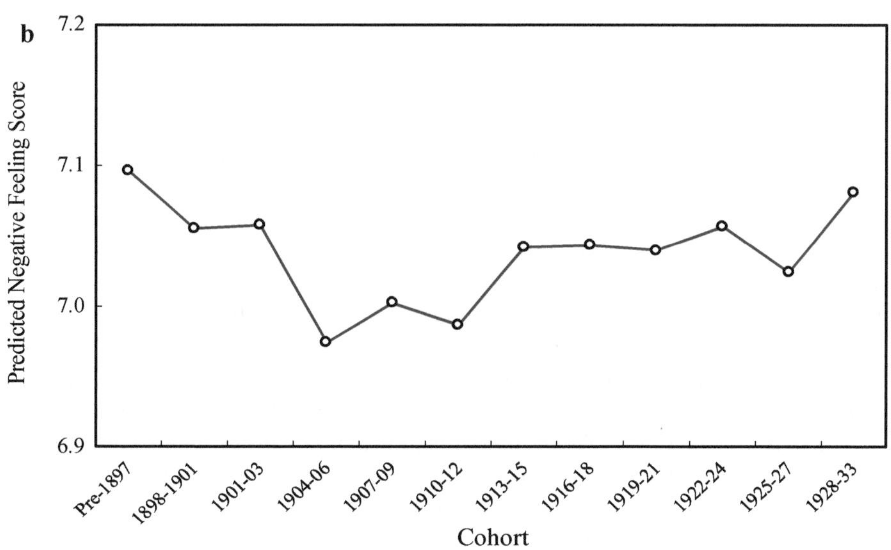

Fig. 4.10b Predicted negative feeling score over birth cohort. (Sources: 1998–2018 Chinese Longitudinal Healthy Longevity Survey (CLHLS))

trend over time period. The cohort effects demonstrated in Fig. 4.7b exhibit a downward trend for earlier birth cohorts. An upward trend is observed for the 1916 to 1918 and later birth cohorts. Our common understanding is that with China experiencing the transition from deaths being caused primarily by infectious diseases to chronic diseases, the number of chronic diseases reported by the oldest old should rise across birth cohorts. Findings of this research, nonetheless, demonstrate a curving pattern instead of a rising trend, which is worth further investigating. It is important to know what social factors have caused such a pattern and how those factors may have played a role in determining Chinese oldest old's chronic disease changes.

Figures 4.8a and 4.8b display the period and cohort effects for predicted cognitive function scores of Chinese oldest old. According to Fig. 4.8a, the predicted cognitive function scores range from 20.5 to 22.2. The period effects show a downward trend during 1998 to 2000. Since year 2000, a general upward trend is observed over time although fluctuations are shown in some years. The magnitude of such period changes is relatively small since the predicted cognitive function scores fall between 20.5 to 22.2 during the studied time periods. Overall, the results suggest that the predicted cognitive function of the oldest old is getting better over time.

As to the cohort variations, Fig. 4.8b indicates that there is an increasing trend in predicted cognitive function before the 1901–03 birth cohort, a decreasing trend is observed afterwards. It is interesting to explore why a general downward trend is found for later birth cohorts.

Figures 4.9a and 4.9b show period and cohort effects when predicting positive feeling scores. The predicted positive feeling scores change from 9.9 in 2008 to 10.6 in 1998 and 2011. The most intriguing finding is the dip in the predicted positive feeling score for year 2008. That was the year when world countries, including.

China, experienced an economic hardship. In terms of the cohort effects, the predicted positive feeling scores range from 10.3 for the birth cohort that was born before 1897. Such scores increase to 10.5 for the 1898 to 1900 birth cohort and then the scores begin to decline among later birth cohorts. A rebound is shown for the 1925–27 birth cohort. These findings indicate that the predicted positive feeling scores are fairly stable over time except for the year that China experienced an economic recession and the *Wenchuan* earthquake. The 1898 to 1900 and 1925–27 birth cohort seem to be the ones that report better positive feelings than other birth cohorts. The predicted positive feeling scores tend to decline across birth cohorts, suggesting younger birth cohorts tend to be less positive as compared to earlier birth cohorts.

The period and cohort effects in predicted negative feeling scores are presented in Figs. 4.10a and 4.10b, respectively. The predicted negative feeling scores vary from 7.8 in 1998 to 6.0 in 2000. In general, a declining trend is observed in terms of the predicted negative feeling scores, indicating Chinese oldest old's subjective wellbeing that is measured by negative feeling scores is getting better over time. The predicted negative feeling scores show a curving trend across birth cohorts with the 1904–06 birth cohort showing the lowest predicted negative feeling scores. The magnitude of cohort changes is relatively small since the predicted negative feeling scores fall between 6.9 to 7.1 over the studied birth cohorts. In short, the results reveal that the predicted negative feeling scores of Chinese oldest old is declining over time, indicating better subjective wellbeing of the oldest old over time. Although the cohort effects are shown, the magnitude of cohort changes is relatively small.

4.4 Conclusion and Discussion

This chapter has showed descriptive results of health measures and demonstrated how Chinese oldest old's health outcomes have changed by time period and across birth cohorts. In addition, the chapter has also investigated the period and cohort variations when only including age, age-squared term, period and birth cohort in CCREMs to predict the oldest old's health. Several important findings emerge from this part of the analyses.

According to descriptive results, SRH and negative feeling scores show a declining trend by period; but cognitive function and the average number of chronic diseases have showed a rising trend over time period. When using age, age-squared term, cohort and period to predict health outcomes, the results show that age-squared term is not significant in CCREMs, suggesting that age linearly relates to health measures in this study. Basically, age has a positive effect on SRH but it negatively relates to other health measures. In this sense, the "age as maturity" hypothesis suggested by the role theory of aging (Yang, 2008) is supported by this research. It makes sense because cognitive function declines with age. The number of chronic diseases is also found to decrease with age. This is probably because those unhealthy individuals died with age increasing and the oldest old who are still alive are most likely to be healthier individuals and may have fewer chronic diseases. Thus, a negative correlation is shown between the number of chronic diseases and age. This finding fells in line with the selective survival theory.

In addition to the above results, significant period and cohort effects are found when only age, age-squared term, period and cohort are included in CCREMs to predict health. The findings suggest that the period variations in health measures are largely consistent with the results drawn from descriptive analyses. The most significant period effect is shown when predicting positive feeling scores of the oldest old. There is a dip in year 2008 when estimating the respondent's positive feeling scores. This is likely to be caused by an economic hardship and the *Wenchuan* earthquake that occurred in that particular year. It implies that unlike physical health, the oldest old's subjective wellbeing is more likely to be influenced by certain social events occurred in unique time periods.

The cohort effects are demonstrated clearly by running CCREMs than by showing results from descriptive analyses. A bell curve is found in predicted SRH across birth cohorts with the 1910–12 birth cohort showing the best SRH. A V-shape trend is observed in predicted numbers of chronic diseases with the 1916–18 birth cohort showing the lowest number of chronic diseases. Such a number rises among younger birth cohorts. A general downward trend is found in cognitive function but an upward trend is revealed in subjective wellbeing across birth cohorts.

In conclusion, this chapter has attempted to investigate how Chinese oldest old's health status has changed over time and across birth cohorts. The current research contributes to existing studies by applying a longitudinal scope to explore how age, period and cohort have simultaneously impacted health of Chinese oldest old. These findings emphasize that Chinese oldest old's health has certainly been affected by period and cohort variations. It is essential to examine the compound effects of age, period and cohorts when detecting the links between the oldest old's health and a variety of demographic, socioeconomic and behavioral factors. The study will attempt to do so in following chapters.

References

Chen, F. N., & Short, S. E. (2008). Household context and subjective well-being among the oldest old in China. *Journal of Family Issues, 29*(10), 1379–1403.

Cronbach, L. J. (1951). Coefficient alpha and the internal structure of tests. *Psychometrika, 16*, 151–160.

Folstein, M. F., Folstein, S. E., & McHugh, P. R. (1975). "Mini-mentalstate": A practical method for grading the cognitive state of patients for the clinician. *Journal of Psychiatric Research, 12*, 189–198.

Lv, X., Li, W., Ma, Y., Chen, H., Zeng, Y., Yu, X., Hofman, A., & Wang, H. (2019). Cognitive decline and mortality among community-dwelling Chinese older people. *BMC Medicine, 17*, 63.

Oftedal, S., Kolt, G. S., Holliday, E. G., Stamatakis, E., Vandelanotte, C., Brown, W. J., & Duncan, M. J. (2019). Associations of health-behavior patterns, mental health and self-rated health. *Preventive Medicine, 118*, 295–303.

Smith, K., & Victor, C. (2019). Typologies of loneliness, living alone and social isolation, and their associations with physical and mental health. *Aging & Society, 39*, 1709–1730.

Yang, Y. (2008). Social inequalities in happiness in the United States, 1972 to 2004: An age-period-cohort analysis. *American Sociological Review, 73*(2), 185–356.

Yang, Y., & Land, K. C. (2006). A mixed models approach to the age-period-cohort analysis of repeated cross-section surveys, with an application to data on trends in verbal test scores. *Sociological Methodology, 36*(1), 75–97.

Yuan, H. (2021). Internet use and mental health problems among older people in Shanghai, China: The moderating roles of chronic diseases and household income. *Aging & Mental Health, 25*(4), 657–663.

Zeng, Y., Vaupel, J., Xiao, Z., Zhang, C., & Liu, Y. (2002). Sociodemographic and health profiles of the oldest old in China. *Population and Development Review, 28*, 251–273.

Demographic Characteristics and the Oldest Old's Health

5

Chapter 4 has described the changing patterns of Chinese oldest old's health by time period and across birth cohorts. Meanwhile, it has demonstrated that Chinese oldest old's health status is affected by age, period and birth cohorts simultaneously. Under this proceeding, chapters 5 through 9 further examine the period and cohort effects when linking a variety of demographic, socioeconomic and behavioral factors to Chinese older adults' health. This chapter focuses on studying the association between demographic characteristics and the oldest old's health. The chapter pays a particular attention to period and cohort effects in the above linkage.

5.1 The Association Between Demographic Characteristics and Older Adults' Health

Demographic factors have been found to have significant impacts on older adults' health. Prior studies have consistently documented that sex is correlated with health of older adults. A group of analyses showed a better health status among males than among females in older ages (Xiong & Yang, 2016; Yang & Zhang, 2019; Yin et al., 2021; Zhang et al., 2021). For instance, a study found that the proportion of male older adults with rheumatoid arthritis was 9.5%, but such a proportion was 15.3% among their female counterparts, indicating a lower disease prevalence among male elders (Li et al., 2013). Researchers also indicated that as compared to males, female older adults are more susceptible to psychological sensitivity and they are more likely to experience depressive symptoms when they are affected by unexpected life events (Wu et al., 2022).This is probably because the average lifespan of females is longer than that of males, resulting in a longer duration of illnesses and a higher prevalence of chronic diseases among females (Zhang et al., 2005).

L. Zhang, *Living Longer and Healthier at Older Ages*, Advances in Studies of Aging and Health 3, https://doi.org/10.1007/978-3-031-69773-9_5

Other scholars, however, challenged the above argument by stating no significant health disparities among male and female older adults. For example, when studying mental health status of older adults with the same background, after controlling for factors such as income, educational attainments and marital status, the sex differences in mental health status among older adults disappeared (Tang & Chen, 2011). Scholars contended that according to the "exposure to risk factors" explanation, psychological health problems of female older adults are not solely caused by sex but by other risk factors. Once other risk factors are controlled, the sex effect diminishes or disappears (Chen et al., 2020; Tang & Chen, 2011; Zhang et al., 2023).

In addition to sex, researchers have also documented a significant effect of age on older adults' health. As age increasing, the risk of having chronic diseases rises, including diabetes, heart disease, cardiovascular diseases et al. This is particularly evident among individuals aged 65 and above who demonstrate a significantly higher prevalence of chronic diseases. Meanwhile, the function of immune system declines with age, making older adults more vulnerable to diseases (Sun & Zhang, 2013; Zhang, 2018). Besides physical health, mental health status is also affected by age. Prior studies showed that the prevalence of depression increases with age. The screening and treatment needs for depression among older adults also become higher as they age (Geldsetzer et al., 2019; Kohler et al., 2017).

Residential factors impact older adults' health as well. Given the rural-urban divide in China, older adults living in urban areas generally reported better health status which is measured by daily functioning and self-rated health (SRH) than those living in rural areas (Bai & Cheng, 2019; Zimmer & Kwong, 2004). Even in the U.S., rural residents showed worse health outcomes than their urban counterparts if indicators such as mortality rates, incidence rates, health behaviors and clinical care were considered (Anderson et al., 2015). The health disparities due to residence may be caused by the fact that older adults residing in urban areas often have better access to readily available healthcare resources so that they may receive more timely treatment than their rural counterparts (Bai & Cheng, 2019). Other scholars, nonetheless, argued that urban older adults only showed health advantages in physical health but not in mental health (Fogelholm et al., 2006). This is because urban older adults experience a faster-paced lifestyle and may face more psychological stress and anxiety than rural dwellers. They are thus more likely to experience depressive symptoms than their rural counterparts (Cho & Cho, 2022; Fang et al., 2015; Nakanishi et al., 2023).

Marital status and widowhood duration have also been found to be influential on older adults' health (Hossain & James, 2022; Krochalk et al., 2008; Perkins et al., 2016). Findings suggested that married elders tended to have better health outcomes as compared to unmarried ones (Hossain & James, 2022). This is because in older ages, losing a spouse may yield a stronger sense of loneliness. Those elders who have lost a spouse may have more difficulties in emotional release, which can lead to severe mental health problems and other types of diseases (Ho, 2018; Xiu et al., 2020). Prior studies indicated that the correlation between marital status and health can be moderated by a variety of factors, including personal characteristics, marital duration, intergenerational relations et al. (Saha et al., 2022). The existing studies, however, are largely cross-sectional. Little research has been done to explore the period and cohort effects in the correlation between demographic characteristics and older adults' health. This chapter intends to fill the voids of prior literature by exploring the period and cohort effects in the study.

5.2 Data, Measures and Methods

5.2.1 Data

This chapter relies on analyzing eight (1998, 2000, 2002, 2005, 2008, 2011–12, 2014 and 2018) waves of the Chinese Longitudinal Healthy Longevity Survey (CLHLS) to constructed CCREMs. The CLHLS has been considered as the first large survey of the oldest old conducted in a developing country (Y. Zeng et al., 2002). The data were collected by Peking University's Center for Healthy Aging and the Family Studies and the China National Research Center on Aging, with support from the U.S. National Institute on Aging. The CLHLS was undertaken in 631 randomly selected counties and cities of 22 provinces and municipalities directly under the Central Government in China. These 22 provinces and municipalities include Liaoning, Jilin, Heilongjiang, Hebei, Beijing, Tianjin, Shanxi, Shaanxi, Shanghai, Jiangsu, Zhejiang, Anhui, Fujian, Jiangxi, Shandong, Henan, Hubei, Hunan, Guangdong, Guangxi, Sichuan, and Chongqing. The survey covers roughly half the counties and cities of those provinces, and the sample areas represent 85.0% of the total population of China.

The 1998 baseline survey includes 9,073 oldest old individuals aged 80 and older. In 2000, 4,844 of those elders were reinterviewed, and 6,372 elders were newly added interviewees. For more detailed description of the data, see Zeng et al. (2002). This chapter focuses on studying older adults aged 85 to 105 since prior literature considered persons who reported age 106 or higher as invalid cases (Y Zeng et al., 2002).

5.2.2 Measures

The measures of demographic characteristics in this part of the analysis include age, sex, residence and marital status. The variable *age* is coded as a continuous variable, ranging from 85 to 105. For the sex variables, males are coded as "1" and females as "0". *Residence* is coded as a dummy variable with those living in cities being coded as "1" and "0" if living in rural areas. *Marriage* is often measured by current marital status of the respondent, that is, by placing the respondent into one of the following categories: married, never married (single), divorced/separate, and widowed. In this research, considering the respondents are 85 and older, simpler classification may be more suitable. Thus, the respondents are classified into two categories. Those who were married are coded as "1" and "0" if otherwise.

Health outcomes are measured in the same way that was described in chap. 4. The health measures include self-rated health (SRH), number of chronic diseases, cognitive function, and positive and negative feelings which are considered are measures of one's subjective wellbeing (Lv et al., 2019; Oftedal et al., 2019; Smith & Victor, 2019; Yuan, 2021).

5.2.3 Methods

In this chapter, a cross-classified random-effects model (CCREM) is used to take into account the embeddedness of respondents in the CLHLS surveys within a time period (Yang & Land, 2006). Specifically, the model estimate fixed effects of individual-level covariates and random effects of period (or time). Individuals are the level-1 unit of analysis. Periods (or time) become the level-2 units of analysis. In the analysis of the CLHLS data, an individual respondent is nested in and cross-classified by a higher level context, which are period (or time) and birth cohort.

Beyond providing random period effects, CCREMs allow for random effects of independent variables, which is used to test for cross-period variations in the effects of key independent variables, i.e. demographic variables. Period-specific random-effects coefficients and cohort-specific random-effects coefficients suggest potential across-period changes and the across-cohort changes in the effects of demographic characteristics on the oldest old's health status, respectively. This study specifies models of the oldest old's health status as a quadratic function of age.

Several models are presented in the analysis. The first model only includes age and age-squared as fixed effects. This model measures the overall effects of period on the oldest old's health status, controlling for age and age squared variables. The next models include demographic measures, controlling for the effects of age and age squared term. The full individual level or level-1 equation can be expressed as:

$$\mathrm{Y_{ijk}} = \beta_{0\mathrm{jk}} + \beta_{1\mathrm{jk}}\mathrm{AGE} + \beta_{2\mathrm{jk}}\mathrm{AGE}^2 + \beta_{3\mathrm{jk}}\mathrm{X} + \sum_{p=10}^{p} \beta_{\mathrm{p}}\mathrm{Xp} + \mathrm{e_{ijk}} \tag{5.1}$$

where $\mathrm{Y_{ijk}}$ represents health status of the ith respondent for i = 1, …, nj individuals within the jth period for j = 1, …. β_{0j} is the intercept or for respondents in period j; "AGE" and "AGE^2" symbolize age and age-squared, respectively. "X" indicates independent variables; i.e., demographic variables. β_p represents other individual-level fixed effects where k is the maximum number of covariates. e_{ij} represents an individual level random error term.

The level-2 model can be expressed as:

$$\beta_{0j} = \gamma_0 + p_{0j} + c_{0k} \tag{5.2}$$

This model specifies the overall mean vary from period to period and from cohort to cohort. γ_0 is the model intercept, which is the expected mean at the zero value of all level-1 variables averaged over all periods and cohorts. Here p_{0j} and c_{0k} are the residual random effects of period and cohort, respectively. The cell mean, $\beta_{0j,}$ is equal to the sum of the overall mean or intercept (γ_0), and the residual random effect of period j (p_{0j}) and the residual random effect of cohort (c_{0k}). The residual random effects of period j (p_{0j}) and cohort k (c_{0k}) examine the effects of each period and cohort on a specific health measure. When adding (2) to (1), it yields the level-1 and level-2 combined model:

$$Y_{ijk} = \gamma_0 + \beta_{0jk} + \beta_{1jk}AGE + \beta_{2jk}AGE^2 + \beta_{3jk}X + p_{0j} + c_{0k} + e_{ijk} \quad (5.3)$$

where, within each survey year j, respondent i's health is modeled as a function of his or her age, age-squared and other control variables. In the above model, γ_0 represents the fixed-effects coefficients for demographic variables. The symbols of p_{0j} and c_{0k} represent the period-specific and cohort-specific effects of demographic variables, respectively. The model tests for random effects of demographic factors on the respondent's health across periods and cohorts. All continuous variables are sorted and centered around their overall mean.

Since there are eight waves of survey data, survey years for the eight waves are considered as the eight periods. I have developed 13 birth cohorts in the analyses (pre-1897, 1898–1900, 1901–03, 1904–06, 1907–09, 1910–12, 1913–15, 1916–18, 1919–21, 1922–24, 1925–27, 1928–30 and 1931–33). The 1934 and later birth cohorts are dropped from the analyses since those cohorts do not fall into the age range of 85 to 105. When showing descriptive results, all 13 birth cohorts are shown in graphs. When conducting regression analyses, the 1928–30 and 1931–33 birth cohorts are combined into one birth cohort considering the number of respondents who were born in and after 1928 is limited.

5.3 Results

5.3.1 Descriptive Statistics

5.3.1.1 Basic Statistics of Variables

Table 4.1 in Chap. 4 has presented descriptive information for all variables used in this research, including the respondent's demographic characteristics. To refresh readers' memory, this part briefly highlights basic descriptive results of the respondent's demographic profiles. Again, since CLHLS is a sampling survey and this research does not use the sampling weight, the percentage distributions refer to the sample. In total, 58,485 oldest old have been surveyed in eight waves of CLHLS. In terms of the demographic characteristics, the mean age of the oldest old is 94.3 with 47.0% of respondents being urban and 53.0% of them being rural. About 13.9% of the oldest old were married when surveys were conducted.

5.3.1.2 Period and Cohort Variations in Demographic Profiles of the Oldest Old

I now use graphs to describe period and cohort variations in demographic characteristics of the oldest old. Figure 5.1 shows how the percentage distributions of male respondents in the overall sample have changed over time period and across birth cohorts. Since the percentage shares of females show the opposite patterns, the

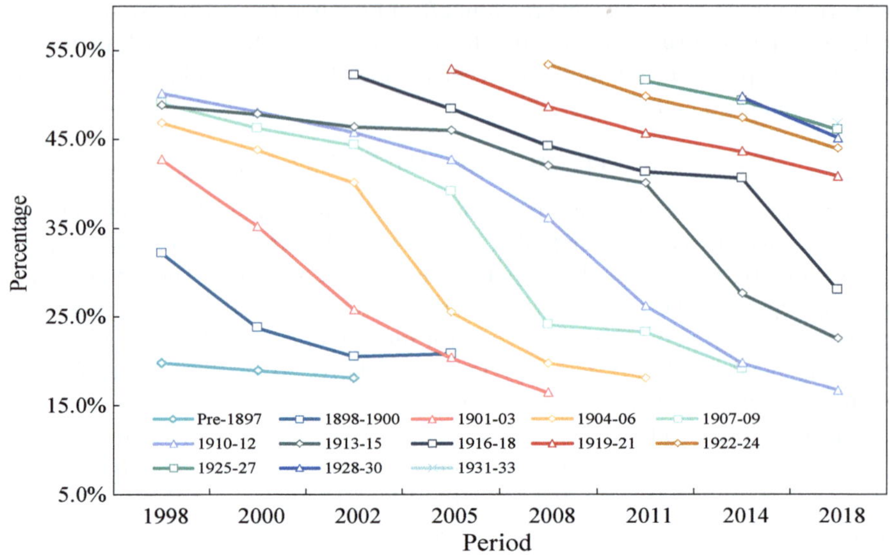

Fig. 5.1 Percentage of Males by Time Period and across Birth Cohorts (Unweighted). (Sources: 1998–2018 Chinese Longitudinal Healthy Longevity Survey (CLHLS))

graph for females is not shown here. Figure 5.1 indicates that the percentage of male respondents in the studied sample has been increasing across birth cohorts. For example, for birth cohort born prior to 1897, males were only about 20.0% of the overall sample. Among the most recent birth cohorts, the percentage of males rose to about half of the total respondents. Period wise, except for the 1928–33 birth cohort, there is a declining trend regarding percentage shares of male respondents for each birth cohort over time. The findings suggest that the percentage shares of male respondents in the overall sample has been increasing among younger birth cohorts. But for a specific birth cohort, the percentage share of male oldest old has declined over time. This makes sense because females live longer than males. Over time, the survival advantage of females has resulted in the percentage shares of females increasing in older ages.

Figure 5.2 shows the percentage distributions of respondents who lived in urban areas. Clearly, the period variations are more pronounced than cohort variations. The percentage shares of urban oldest old range from 27.8% in 1998 to 64.0% in 2000 and 61.4% in 2018 of the total sample. The 2000, 2011 and 2018 waves have shown higher percentage shares of urban respondents as compared to other waves. For the rest of the five waves, the percentage shares of urban residents were lower than 50.0%. Since residence is coded as a dummy variable, the cohort and period variations for rural respondents show the opposite patterns as compared to the urban respondents. Thus, the figure for rural respondents is not shown here.

The period and cohort variations in terms of the oldest old's marital status are depicted by Fig. 5.3. The results show that the percentages of respondents who claimed themselves as being married are in an increasing trend across birth cohorts.

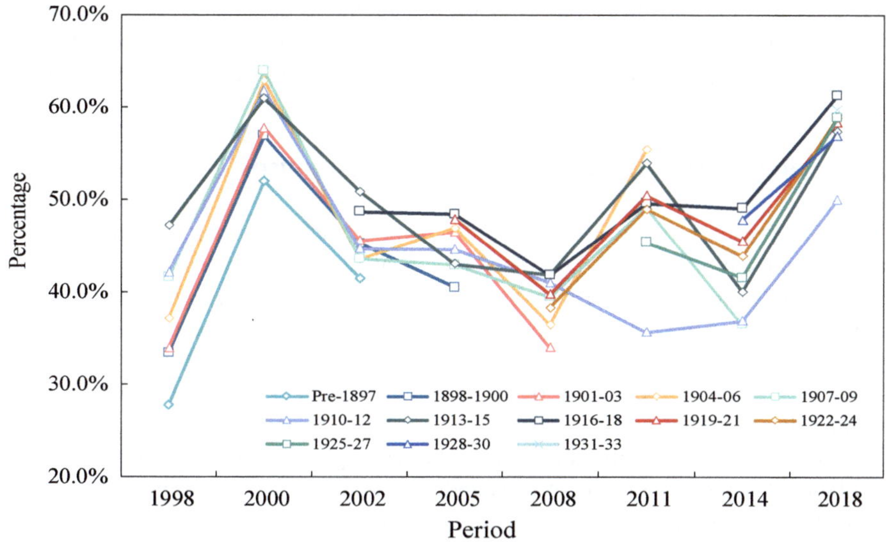

Fig. 5.2 Percentage of Urban Respondents by Time Period and across Birth Cohorts (Unweighted). (Sources: 1998–2018 Chinese Longitudinal Healthy Longevity Survey (CLHLS))

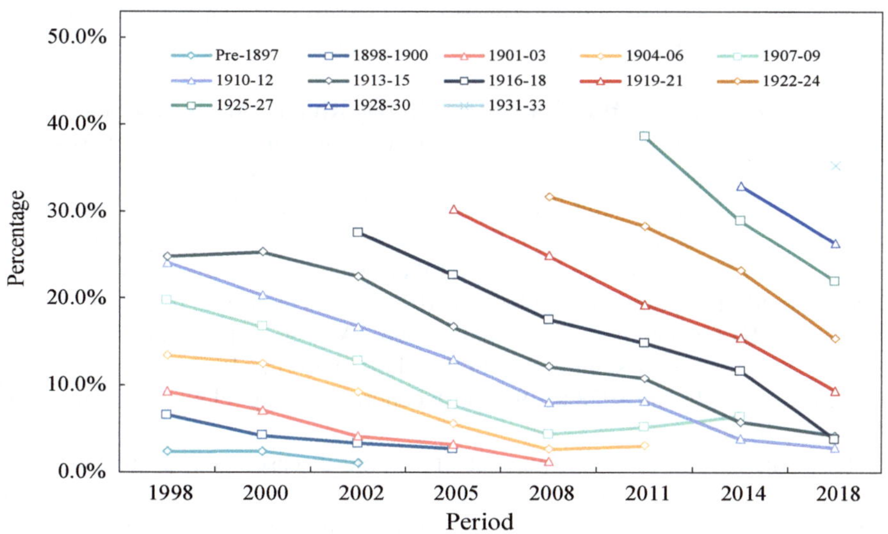

Fig. 5.3 Percentage of Married Respondents by Time Period and across Birth Cohorts (Unweighted). (Sources: 1998–2018 Chinese Longitudinal Healthy Longevity Survey (CLHLS))

But for each specific birth cohort, such a percentage declines over time. This is because as time goes by, some respondents have lost their spouses and became widowed. Thus, the percentages of married respondents decrease for each birth cohort.

5.3.2 Regression Results Based on APC Analyses

Table 5.1 presents regression results from the CCREMs when including demographic and socioeconomic variables in the analyses. As expected, men on average report a lower number of chronic diseases and show better SRH, cognitive function and subjective wellbeing as compared to women. For instance, being men increases the predicted SRH, cognitive function, positive feeling scores by 0.07, 1.70 and 0.11 points ($p < .001$), respectively. Being males also reduces the predicted number of chronic diseases and negative feeling scores by 0.09 and 0.24 points ($p < .001$), respectively. Married individuals show better cognitive function and subjective wellbeing, but they report worse SRH and a greater number of chronic diseases than those who were divorced, widowed or separated.

Next, I put each of the demographic variables into the second level of the model to investigate the period and cohort effects when controlling for SES covariates. Tables 5.2, 5.3, and 5.4 show regression results from the CCREMs when include sex, residence, marital status variables in the second level of the model, respectively. Only significant period and cohort variations are presented by graphs. The next several subsections show significant period and cohort variations in the oldest old's health differentials by sex, residence and marital status, respectively.

5.3.2.1 Period and Cohort Variations in Health Disparities by Sex

Table 5.2 shows when sex is included in level 2, whether significant period and cohort changes occur by sex net of the effects of age and other factors. The results indicate that when predicting SRH by sex, significant period and cohort variations exist, which are demonstrated by Figs. 5.4a and 5.4b, respectively. Figure 5.4a shows that men tend to have better SRH than women. Although there is a slight increase in SRH in years 2002, 2008 and 2018, there is an overall declining trend in predicted SRH for both male and female groups. As Fig. 5.4b shows, disparities in SRH by sex remain constant across birth cohorts with the 1904 to 1906, 1913 to 1915, 1922 to 1924 and 1928 to 1933 birth cohorts showing relatively smaller SRH disparities by sex.

As to the second measure, number of chronic diseases, the results show that period and cohort variations in predicted number of chronic diseases by sex are significant, which are depicted by Figs. 5.5a and 5.5b. The figure suggests that overall, the predicted number of chronic diseases has been increasing over time. There is an interesting pattern regarding gender differences, that is, during 1998 to 2005, the predicted number of chronic diseases among male oldest old is lower than that of females. There are crossovers in years 2008 and 2011 and the predicated number of chronic diseases becomes higher for males than that for females afterwards.

Table 5.1 Regression results of health outcome measures on demographic and socioeconomic variables from CCREMs: Chinese Aged 85–105

	Self-rated health		Chronic disease		Cognitive function		Positive feelings		Negative feelings	
Variables	Coef.	P Value	Coef.	P Value	Coef.	P Value	Coef.	P Value	Coef.	P Value
Fixed effect component										
Intercept	3.306	<0.001	0.873	<0.001	19.317	<0.001	9.861	<0.001	7.691	<0.001
Control variables										
Age	0.002	0.039	−0.014	<0.001	−0.350	<0.001	−0.016	<0.001	−0.005	0.032
Age^2	−0.000	0.565	0.000	0.083	0.001	0.393	0.000	0.532	−0.001	0.191
Sex (male = 1)	0.069	<0.001	−0.085	<0.001	1.705	<0.001	0.108	<0.001	−0.235	<0.001
Residence (urban =1)	0.050	<0.001	0.276	<0.001	0.407	<0.001	0.353	<0.001	−0.294	<0.001
Being married	−0.052	<0.001	0.071	<0.001	0.632	<0.001	0.059	0.076	−0.432	<0.001
Schooling years	0.010	<0.001	0.045	<0.001	0.259	<0.001	0.067	<0.001	−0.048	<0.001
Often going to bed hungry as a child (ref. = hungry)	0.050	<0.001	0.017	0.163	0.614	<0.001	0.184	<0.001	−0.214	<0.001
Random effect component										
Period										
Intercept	0.008	0.034	0.061	0.033	0.407	0.037	0.042	0.032	0.124	0.032
Cohort										
Intercept	0.000	0.113	0.008	0.018	0.104	0.027	0.003	0.113	0.001	0.164
AIC	124,890.5		168,750.0		306,935.7		216,510.2		203,902.7	
BIC	124,884.5		168,744.0		306,929.7		216,504.2		203,896.7	

Source: 1998–2018 Chinese Longitudinal Healthy Longevity Survey (CLHLS)

Table 5.2 Regression results of health outcome measures from CCREMs when sex is included in level 2: Chinese aged 85–105

	Self-Rated health		Chronic disease		Cognitive function		Positive feelings		Negative feelings	
Variables	Coef.	P Value	Coef.	P Value	Coef.	P Value	Coef.	P Value	Coef.	P Value
Fixed effect component										
Intercept	3.415	<0.001	0.926	<0.001	20.694	<0.001	10.263	<0.001	7.240	<0.001
Demographic & SES variables										
Age	0.002	0.082	−0.016	<0.001	−0.366	<0.001	−0.018	<0.001	−0.004	0.075
Age^2	0.000	0.489	0.000	0.122	0.001	0.243	0.000	0.479	0.000	0.272
Sex (male = 1)	0.086	<0.001	−0.018	0.596	2.147	<0.001	0.224	<0.001	−0.318	<0.001
Residence (urban =1)	0.055	<0.001	0.300	<0.001	0.542	<0.001	0.385	<0.001	−0.319	<0.001
Being married	−0.032	0.009	0.081	<0.001	0.701	<0.001	0.061	0.063	−0.444	<0.001
Education level (ref. = primary school)										
Middle school	0.079	<0.001	0.287	<0.001	1.893	<0.001	0.497	<0.001	−0.285	<0.001
Higher education	0.039	0.049	0.487	<0.001	1.662	<0.001	0.482	<0.001	−0.388	<0.001
Often going to bed hungry as a child (ref. = hungry)	−0.060	<0.001	−0.042	<0.001	−0.789	<0.001	−0.225	<0.001	0.242	<0.001
Random effect component										
Period										
Intercept	0.007	0.036	0.066	0.033	0.379	0.040	0.037	0.035	0.126	0.032
Male	0.001	0.239	0.006	0.105	0.103	0.129	0.005	0.141	0	–
Cohort										
Intercept	0.000	0.156	0.009	0.023	0.092	0.034	0.002	0.149	0.001	0.208
Male	0.001	0.166	0.003	0.185	0.279	0.072	0	–	0.001	0.295
AIC	128,755.0		174,805.2		316,186.9		223, 085.6		210,167.9	
BIC	128,745.0		174,795.2		316,176.9		223,077.6		210,159.9	

Source: 1998–2018 Chinese Longitudinal Healthy Longevity Survey (CLHLS)

Table 5.3 Regression results of health outcome measures from CCREMs when residence is included in level 2: Chinese aged 85–105

	Self-Rated health		Chronic disease		Cognitive function		Positive feelings		Negative feelings	
Variables	Coef.	P Value	Coef.	P Value	Coef.	P Value	Coef.	P Value	Coef.	P Value
Fixed effect component										
Intercept	3.413	<0.001	0.922	<0.001	20.717	<0.001	10.269	<0.001	7.238	<0.001
Demographic & SES variables										
Age	0.002	0.060	−0.015	<0.001	−0.355	<0.001	−0.017	<0.001	−0.004	0.066
Age ^{2}	0.000	0.617	0.000	0.115	0.001	0.530	0.000	0.491	−0.001	0.141
Sex (male = 1)	0.088	<0.001	−0.028	0.024	2.112	<0.001	0.223	<0.001	−0.318	<0.001
Residence (urban =1)	0.054	<0.001	0.315	0.001	0.536	<0.001	0.384	<0.001	−0.297	0.002
Being married	−0.033	0.008	0.081	<0.001	0.666	<0.001	0.060	0.067	−0.445	<0.001
Education level (ref. = primary school)										
Middle school	0.079	<0.001	0.285	<0.001	1.876	<0.001	0.498	<0.001	−0.287	<0.001
Higher education	0.039	0.048	0.487	<0.001	1.654	<0.001	0.481	<0.001	−0.394	<0.001
Often going to bed hungry as a child (ref. = hungry)	−0.059	<0.001	−0.042	<0.001	−0.792	<0.001	−0.224	<0.001	0.241	<0.001
Random effect component										
Period										
Intercept	0.008	0.036	0.044	0.034	0.426	0.037	0.041	0.036	0.122	0.032
Urban	0.000	0.453	0.012	0.047	0.010	0.332	0.015	0.067	0.019	0.062
Cohort										
Intercept	0.001	0.108	0.007	0.024	0.112	0.028	0.002	0.130	0	–
Urban	0.000	0.473	0.001	0.226	0.016	0.251	0	–	0.006	0.083
AIC	128,760.2		174,749.2		316,202.7		223,072.2		210,134.8	
BIC	128,750.2		174,739.2		316,192.7		223,064.2		210,126.8	

Source: 1998–2018 Chinese Longitudinal Healthy Longevity Survey (CLHLS)

Table 5.4 Regression results of health outcome measures from CCREMs when marital status is included in level 2: Chinese aged 85–105

	Self-rated health		Chronic disease		Cognitive function		Positive feelings		Negative feelings	
Variables	Coef.	P Value	Coef.	P Value	Coef.	P Value	Coef.	P Value	Coef.	P Value
Fixed effect component										
Intercept	3.416	<0.001	0.928	<0.001	20.709	<0.001	10.262	<0.001	7.241	<0.001
Demographic & SES variables										
Age	0.002	0.052	−0.015	<0.001	−0.357	<0.001	−0.018	<0.001	−0.004	0.080
Age ^{2}	0.000	0.452	0.000	0.080	0.001	0.422	0.000	0.516	0.000	0.202
Sex (male = 1)	0.088	<0.001	−0.025	0.040	2.107	<0.001	0.220	<0.001	−0.318	<0.001
Residence (urban =1)	0.054	<0.001	0.301	<0.001	0.546	<0.001	0.385	<0.001	−0.319	<0.001
Being married	−0.033	0.130	0.086	0.037	0.688	0.002	0.071	0.221	−0.432	<0.001
Education level (ref. = primary school)										
Middle school	0.079	<0.001	0.285	<0.001	1.875	<0.001	0.497	<0.001	−0.288	<0.001
Higher education	0.039	0.051	0.489	<0.001	1.655	<0.001	0.482	<0.001	−0.390	<0.001
Often going to bed hungry as a child (ref. = hungry)	−0.059	<0.001	−0.042	<0.001	−0.791	<0.001	−0.224	<0.001	0.242	<0.001
Random effect component										
Period										
Intercept	0.007	0.035	0.065	0.033	0.441	0.037	0.040	0.042	0.124	0.032
Being married	0.002	0.127	0.006	0.102	0.028	0.319	0.013	0.057	0.017	0.098
Cohort										
Intercept	0.000	0.133	0.008	0.020	0.103	0.027	0.002	0.141	0.002	0.145
Being married	0	–	0.001	0.358	0.068	0.245	0	–	0.000	0.486
AIC	128,753.8		174,808.0		316,202.5		223,084.0		210,162.4	
BIC	128,745.8		174,798.0		316,192.5		223,076.0		210,152.4	

Source: 1998–2018 Chinese Longitudinal Healthy Longevity Survey (CLHLS)

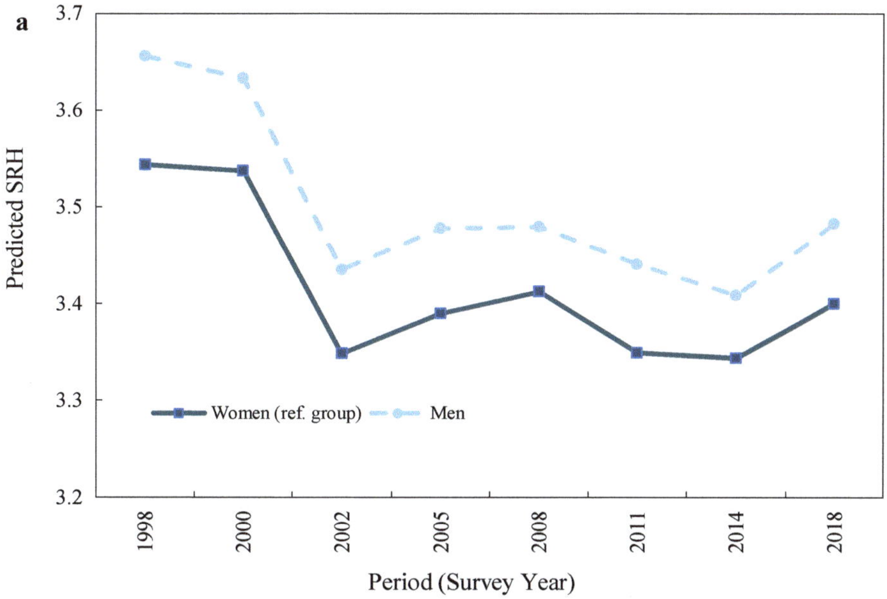

Fig. 5.4a Predicted self-rated health by Sex from CCREMs over time period. (Sources: 1998–2018 Chinese Longitudinal Healthy Longevity Survey (CLHLS))

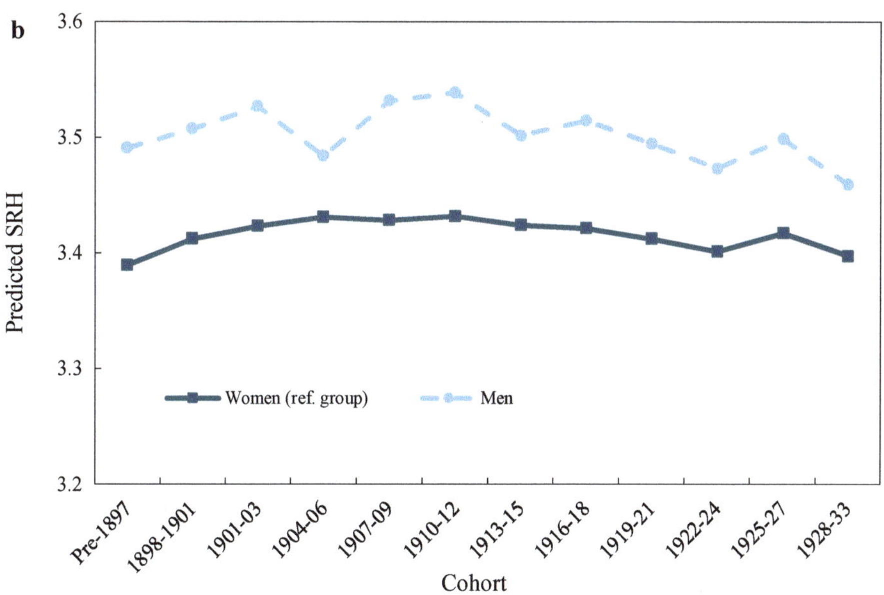

Fig. 5.4b Predicted self-rated health by Sex from CCREMs over birth cohort. (Sources: 1998–2018 Chinese Longitudinal Healthy Longevity Survey (CLHLS))

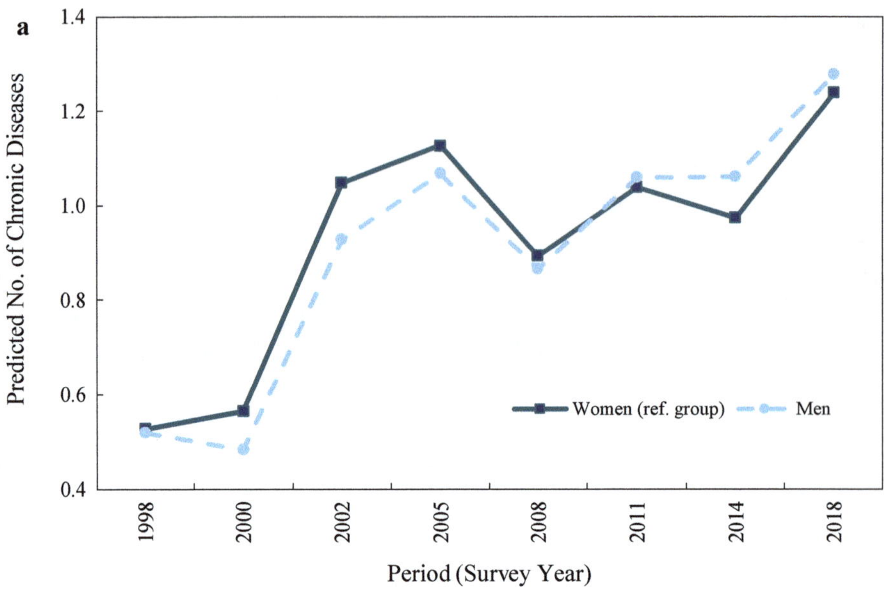

Fig. 5.5a Predicted number of chronic diseases by Sex from CCREMs over time period. (Sources: 1998–2018 Chinese Longitudinal Healthy Longevity Survey (CLHLS))

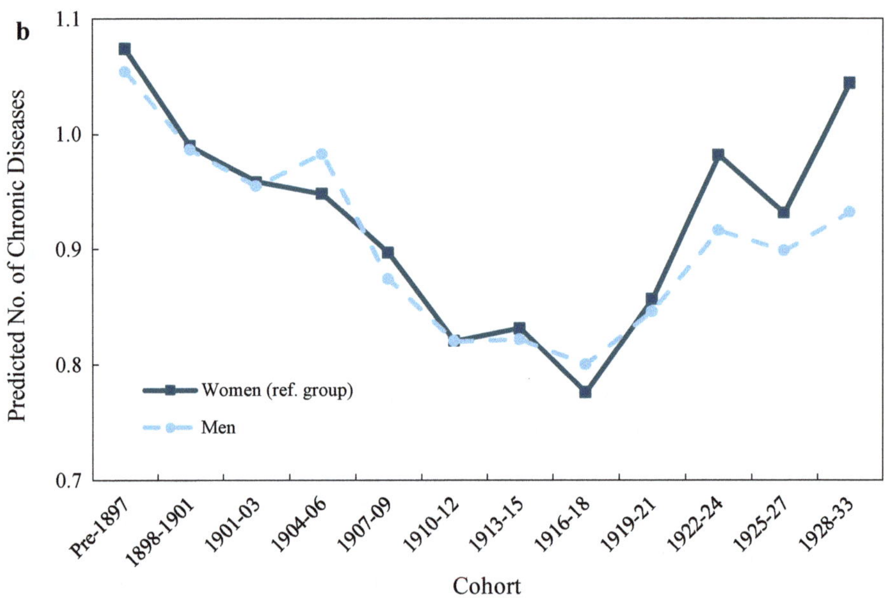

Fig. 5.5b Predicted number of chronic diseases by Sex from CCREMs over birth cohort. (Sources: 1998–2018 Chinese Longitudinal Healthy Longevity Survey (CLHLS))

When cognitive function is considered, results from the CCREMs show significant period and cohort effects, which are described in Figs. 5.6a and 5.6b accordingly. Based on Fig. 5.6a, both male and female predicted cognitive function scores show a general increasing trend over time with fluctuations occurring in some years. Men have higher cognitive function scores than women in all survey years. It seems that since year 2005, the gap in predicted cognitive function by sex enlarge in more recent years, indicating advantages in cognitive function among male than female oldest old. In terms of the cohort effects, Fig. 5.6b indicates that disparities in cognitive function by sex seem to shrink across birth cohorts. It suggests that male advantage in cognitive function tend to be less apparent among more recent birth cohorts.

Regarding the respondents' positive feeling scores, only period shows significant effects. The cohort variations are not significant. Therefore, only significant period effects are depicted by Fig. 5.7. Based on the figure, men show higher positive feeling scores than women in all survey years. The significant period effects are mainly shown in the dint that occurs in year 2008 for both male and female subgroups. As it was stated earlier, the dip may be due to the worldwide economic hardship and the *Wenchuan* earthquake that occurred in that year, which heavily influenced Chinese oldest old's subjective wellbeing.

For the last health measure, negative feeling scores, only significant cohort effects are found based on results from CCREMs, which is depicted by Fig. 5.8. As the figure shows, both male and female negative feeling scores are relatively stable across birth cohorts, with males showing more cohort variations than females.

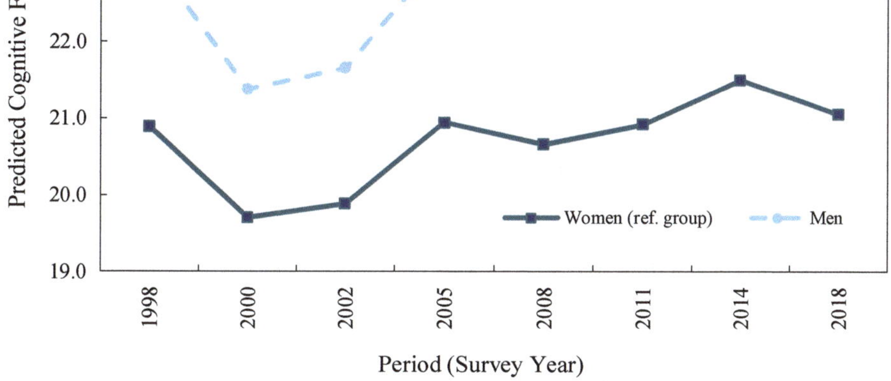

Fig. 5.6a Predicted cognitive function score by sex from CCREMs over time period. (Sources: 1998–2018 Chinese Longitudinal Healthy Longevity Survey (CLHLS))

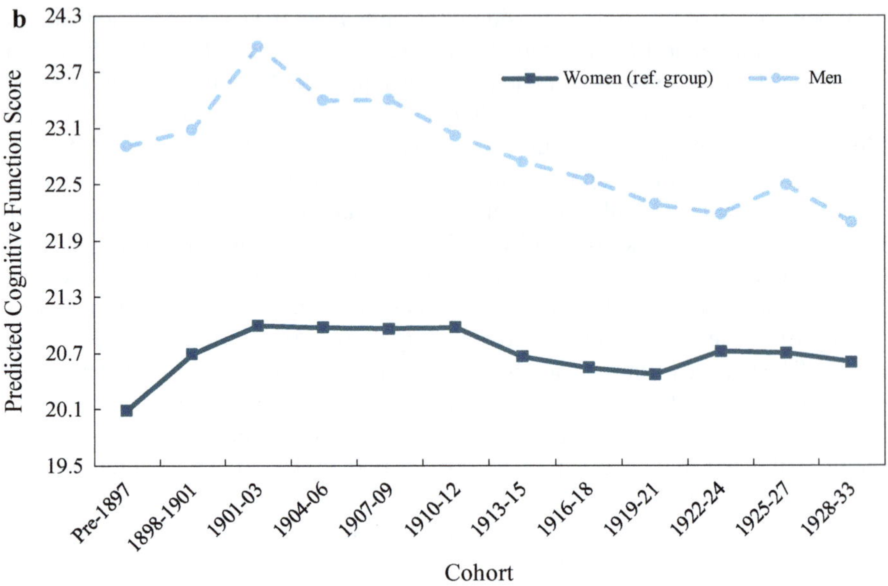

Fig. 5.6b Predicted cognitive function score by sex from CCREMs over birth cohort. (Sources: 1998–2018 Chinese Longitudinal Healthy Longevity Survey (CLHLS))

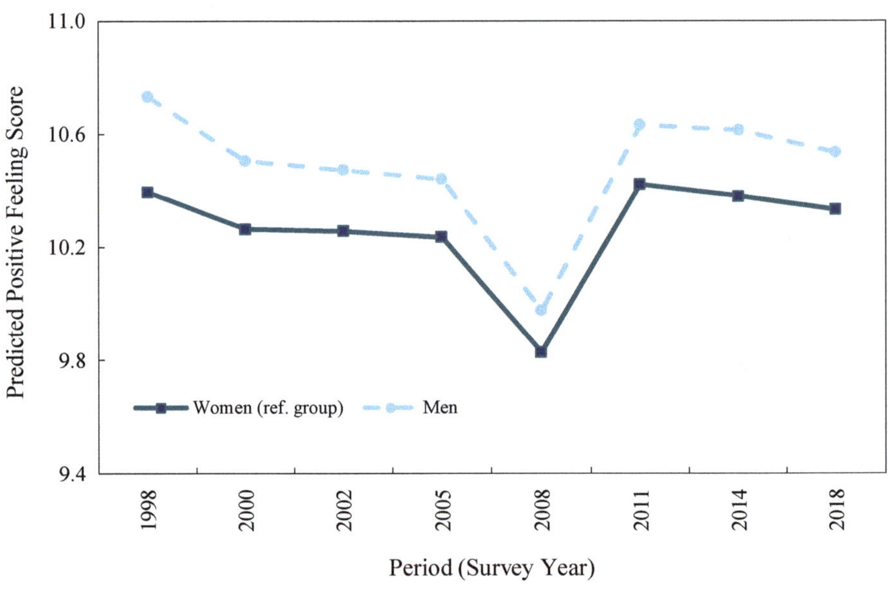

Fig. 5.7 Predicted positive feeling score by sex from CCREMs over time period. (Sources: 1998–2018 Chinese Longitudinal Healthy Longevity Survey (CLHLS))

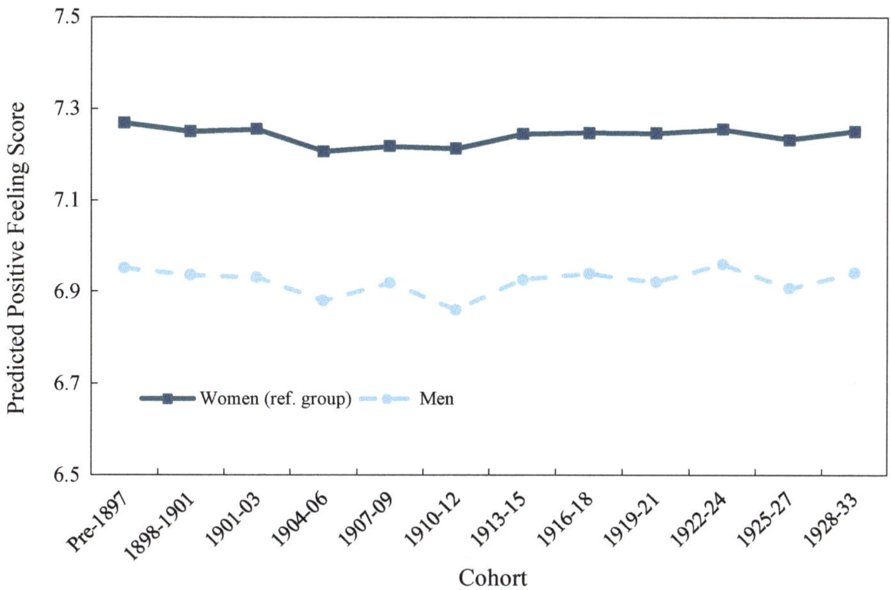

Fig. 5.8 Predicted negative feeling score by sex from CCREMs over birth cohort. (Sources: 1998–2018 Chinese Longitudinal Healthy Longevity Survey (CLHLS))

It suggests that the negative feeling is more stable among women than among men across birth cohorts. In addition, women have higher predicted negative feeling scores than men, which is consistent with the findings for positive feeling scores by sex. It again implies that male oldest old tend to have better subjective wellbeing than their female counterpart. Male advantage is observed once again when it comes to the oldest old's subjective wellbeing.

5.3.2.2 Period and Cohort Variations in Health Disparities by Residence

This section focuses on investigating the period and cohort variations in the association between residence and the oldest old's health, controlling for age and other demographic and SES factors. Figures are generated based on the CCREM results which show significant period and cohort effects. When predicting SRH, the regression models show significant period and cohort effects, which are described in Figs. 5.9a and 5.9b. As it is indicated by both figures, urban oldest old have better predicted SRH as compared to their rural counterparts. The health disparities in predicted SRH by residence remain over time. The significant period effects are shown by a general declining trend in predicted SRH for both rural and urban oldest old with some fluctuations. The predicted SRH seem to be the lowest in years 2002 and 2014. There are slight upward trends in years 2008 and 2018. The cohort effects are shown by a curvy pattern with the 1910–12 birth cohort showing the best SRH.

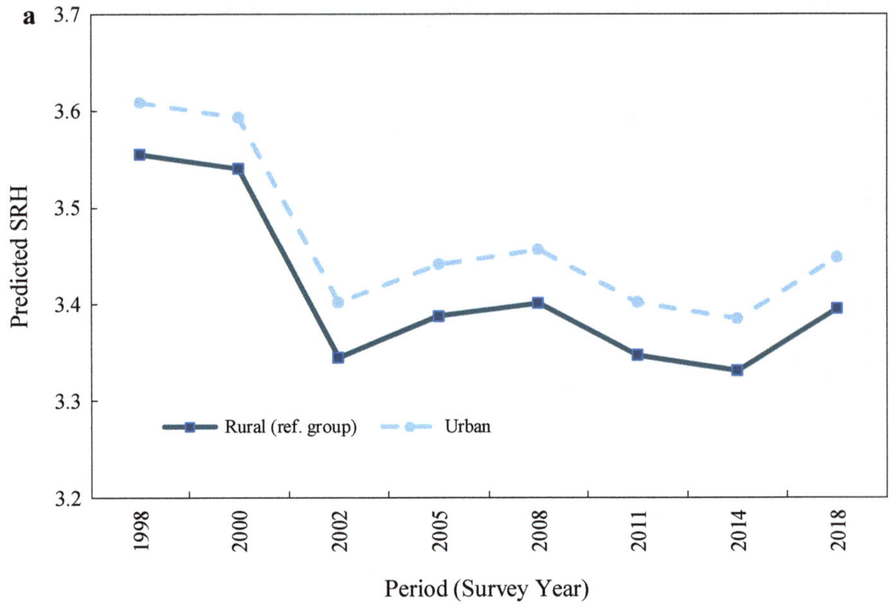

Fig. 5.9a Predicted self-rated health by residence from CCREMs over time period. (Sources: 1998–2018 Chinese Longitudinal Healthy Longevity Survey (CLHLS))

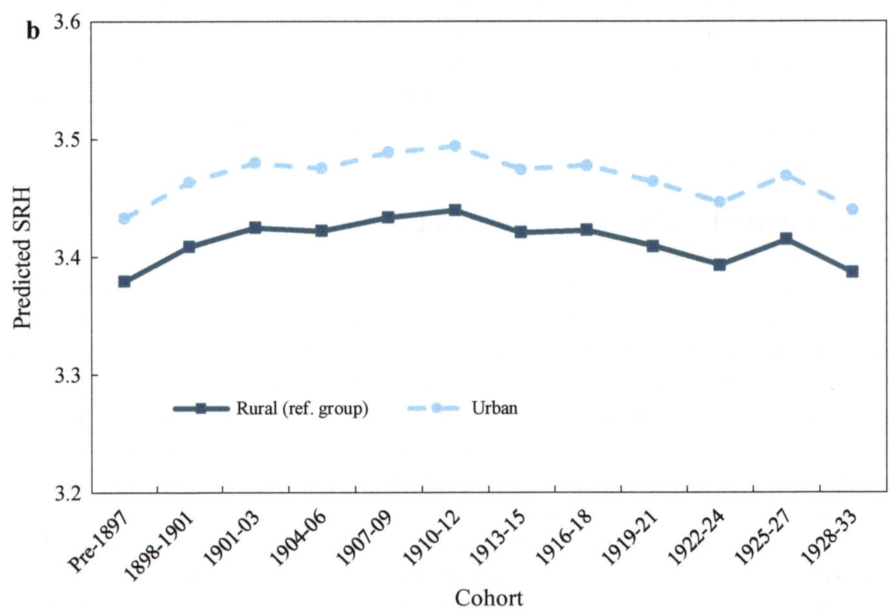

Fig. 5.9b Predicted self-rated health by residence from CCREMs over birth cohort. (Sources: 1998–2018 Chinese Longitudinal Healthy Longevity Survey (CLHLS))

When predicting the second measure of health outcomes, number of chronic diseases, the CCREM results show significant period and cohort effects as well, which are presented by Figs. 5.10a and 5.10b. According to these figures, urban oldest old show a greater predicted number of chronic diseases than rural ones across birth cohorts. Based on Fig. 5.10a, the gaps in urban-rural chronic diseases seem to enlarge over time. The cohort variations are shown by a curvy shape in predicted cognitive function scores for both rural and urban oldest old. That is, the predicted number chronic diseases has been declining for earlier birth cohorts. Starting from the 1916–18 birth cohort, an increasing trend is observed across birth cohorts. These results suggest that chronic diseases among the oldest old first show a declining trend; recent birth cohorts tend to report more chronic diseases than previous birth cohorts.

Figures 5.11a and 5.11b show significant period and cohort effects from CCREMs when predicting cognitive function scores by residence, controlling for age and other demographic and SES covariates. The figures indicate that urban oldest old show better predicted cognitive function scores than their rural counterparts. Figure 5.11a shows that over time, the cognitive function for both rural and urban oldest old is getting better. But the gaps in cognitive function by residence remain over time. Figure 5.11b further depicts that the urban advantage persists across birth cohorts with some cohort fluctuations. Specifically, the 1898–1900 birth cohort shows the best predicted cognitive function. A declining trend is observed afterwards and there is a slight rebound for the 1925–1927 birth cohort. These results highlight that the cognitive function status has been varied by residence and across time period and birth cohorts.

For the last two measures of health outcomes, i.e., positive and negative feeling scores, the CCREM results show significant period effects only, which are displayed in Figs. 5.12 and 5.13, respectively. As Fig. 5.12 shows, the predicted positive feeling score is higher for urban than for rural oldest old. Such scores show a declining trend for both subgroups with a significant dip occurring in year 2008. An increasing trend is observed afterwards. These results indicate that social events are more likely to have a substantial influence on the oldest old's mental health status than physical health condition.

Regarding the predicted negative feeling score, it is higher for rural than for urban oldest old over all periods. The predicted negative feeling scores for urban and rural respondents are initially similar. Then there is a sharp decline in predicted negative feeling scores for both subgroups in year 2000. Fluctuations are shown in later years with period variations being more apparent among the rural oldest old subgroup. This finding suggests that rural oldest old's negative feeling score is more sensitive to period changes as compared to that of their urban counterpart. In year 2018, the predicted negative feeling scores for both subgroups decline, implying that in more recent years the respondents' subjective wellbeing is getting better.

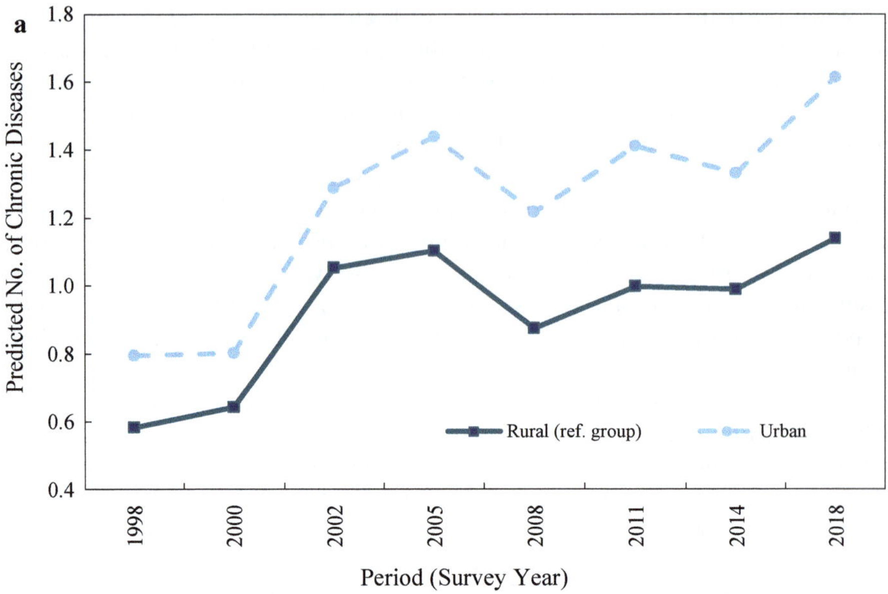

Fig. 5.10a Predicted cognitive function score by residence from CCREMs over time period. (Sources: 1998–2018 Chinese Longitudinal Healthy Longevity Survey (CLHLS))

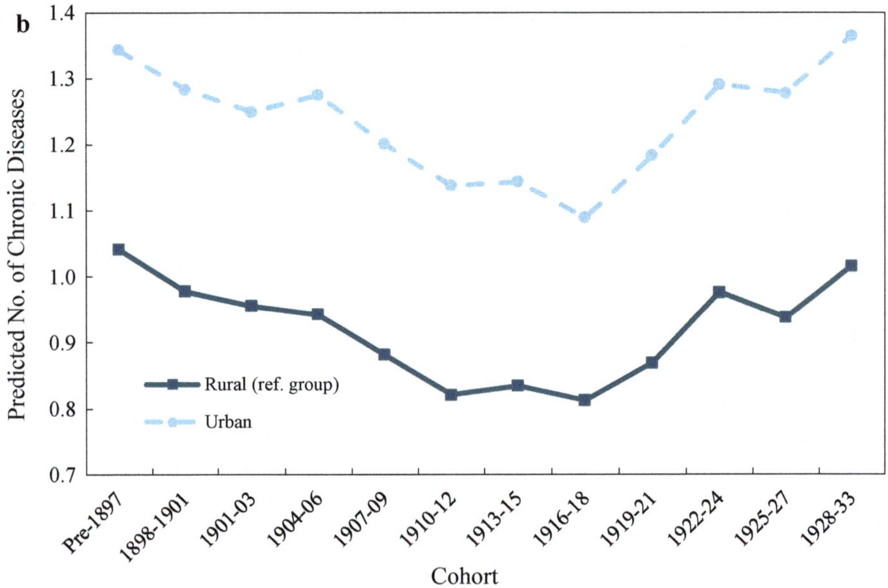

Fig. 5.10b Predicted cognitive function score by residence from CCREMs over birth cohort. (Sources: 1998–2018 Chinese Longitudinal Healthy Longevity Survey (CLHLS))

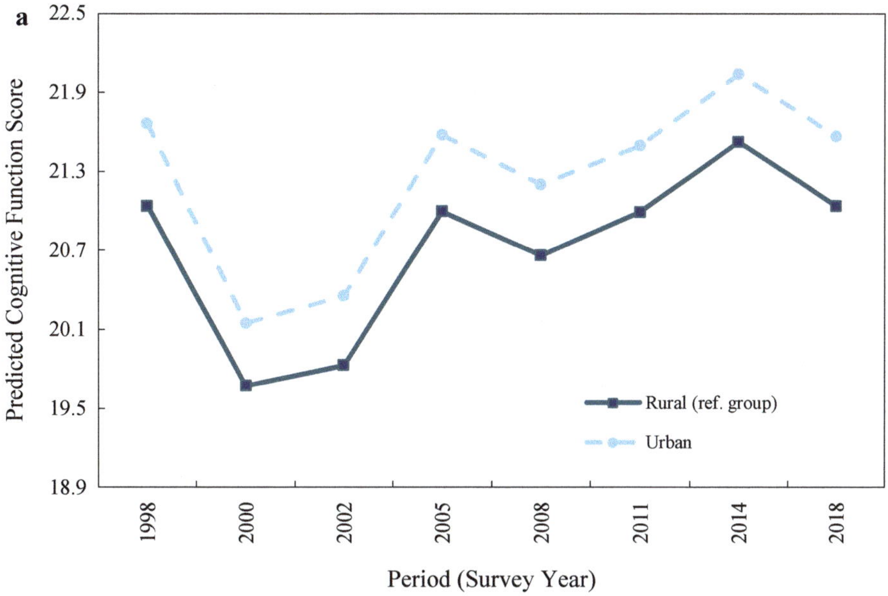

Fig. 5.11a Predicted cognitive function score by residence from CCREMs over time period. (Sources: 1998–2018 Chinese Longitudinal Healthy Longevity Survey (CLHLS))

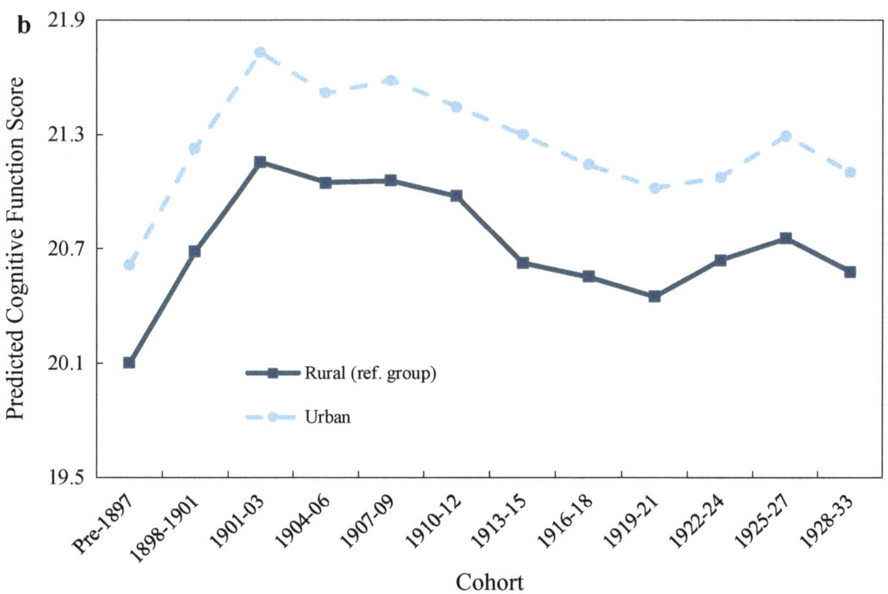

Fig. 5.11b Predicted cognitive function score by residence from CCREMs over birth cohort. (Sources: 1998–2018 Chinese Longitudinal Healthy Longevity Survey (CLHLS))

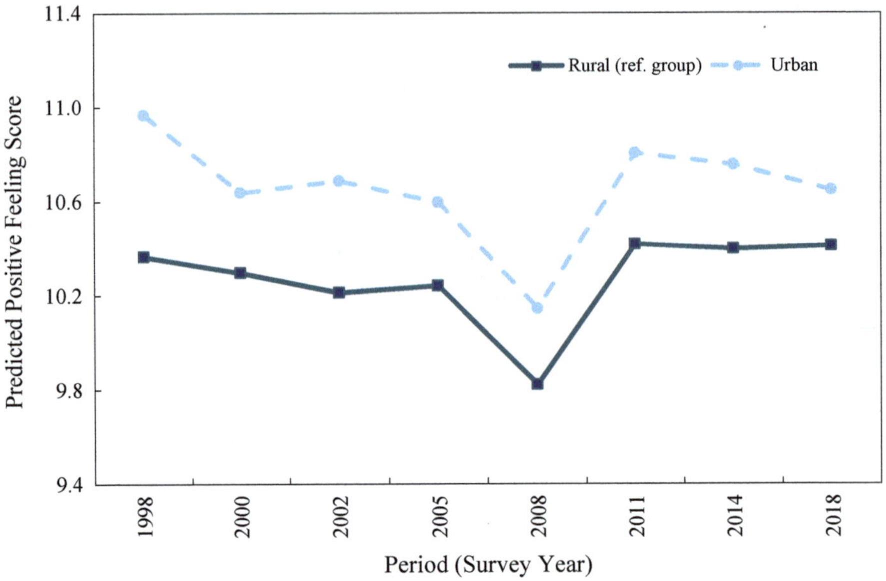

Fig. 5.12 Predicted positive feeling score by residence from CCREMs over time period. (Sources: 1998–2018 Chinese Longitudinal Healthy Longevity Survey (CLHLS))

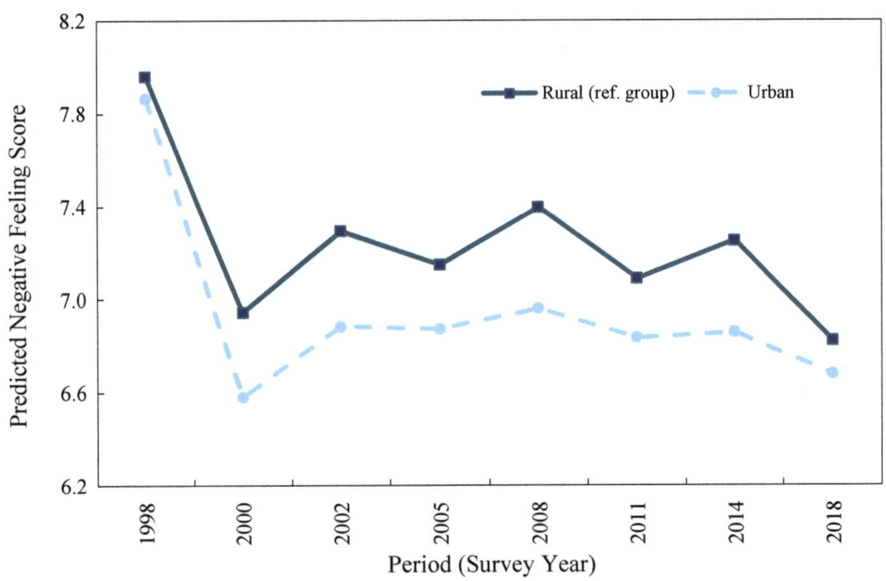

Fig. 5.13 Predicted negative feeling score by residence from CCREMs over time period. (Sources: 1998–2018 Chinese Longitudinal Healthy Longevity Survey (CLHLS))

5.3.2.3 Period and Cohort Variations in Health Disparities by Marital Status

This part of the study investigates the period and cohort variations in the association between marital status and the oldest old's health, controlling for age and other demographic and SES variables. Significant period effects are observed when predicting SRH, which is shown in Fig. 5.14. According to this figure, the predicted SRH for both married and non-married subgroups show a general downward trend with a slight increase in a couple of survey years. The SRH is slightly better for married oldest old in year 1998. Crossovers occur in years 2000 and 2002. Since 2002, the SRH becomes better for non-married than married subgroup. This finding highlights clear period variations, which also challenges our common understanding that married individuals report better health as compared to non-married ones.

Figures 5.15a and 5.15b show predicted number of chronic diseases for the married and non-married subgroups by time period and across birth cohorts, respectively. It seems that the predicted number of chronic diseases for both subgroups has been increasing over time. When demonstrating the period effects, Fig. 5.15a shows that although crossovers occur in a couple of survey years, married oldest old generally has a greater number of predicted chronic diseases than non-married oldest old. In years 2005 and 2011, the health disadvantage of married oldest old, indicated by number of chronic diseases, is more apparent as compared to other survey years. When it comes to the cohort effects, the health disadvantage in chronic diseases among the married group is more apparent. A "V" shape is shown in the predicted number of chronic diseases across birth cohorts. For both married and non-married

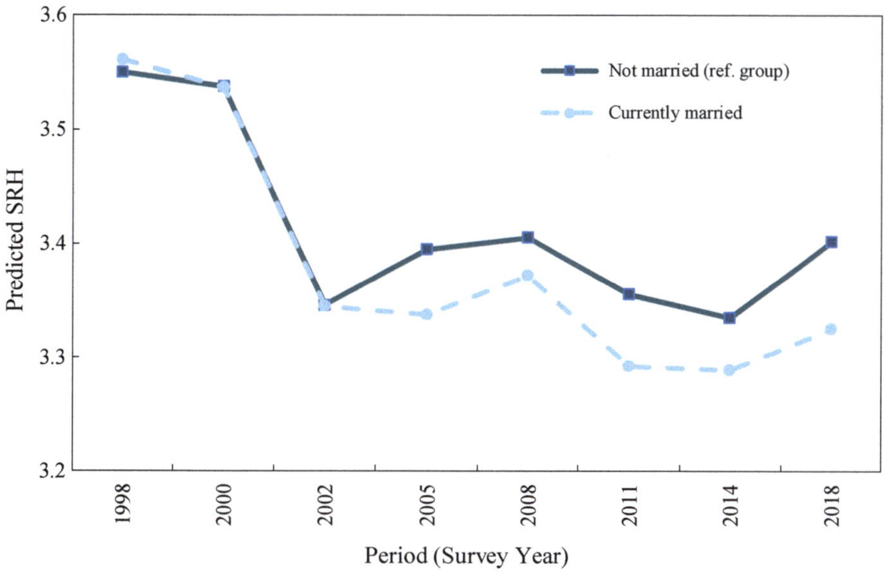

Fig. 5.14 Predicted Self-Rated Health by Marital Status from CCREMs over Time Period. (Sources: 1998–2018 Chinese Longitudinal Healthy Longevity Survey (CLHLS))

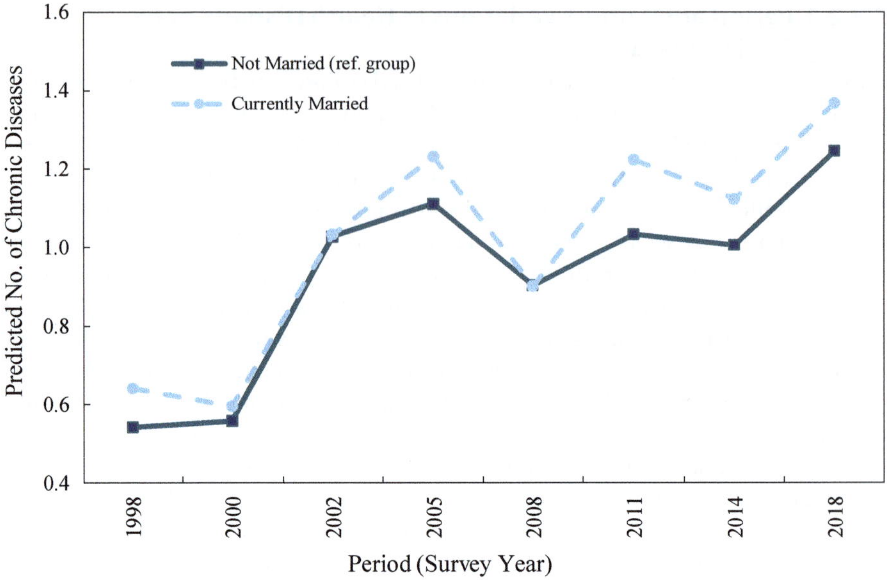

Fig. 5.15a Predicted number of chronic diseases by marital status from CCREMs over time period. (Sources: 1998–2018 Chinese Longitudinal Healthy Longevity Survey (CLHLS))

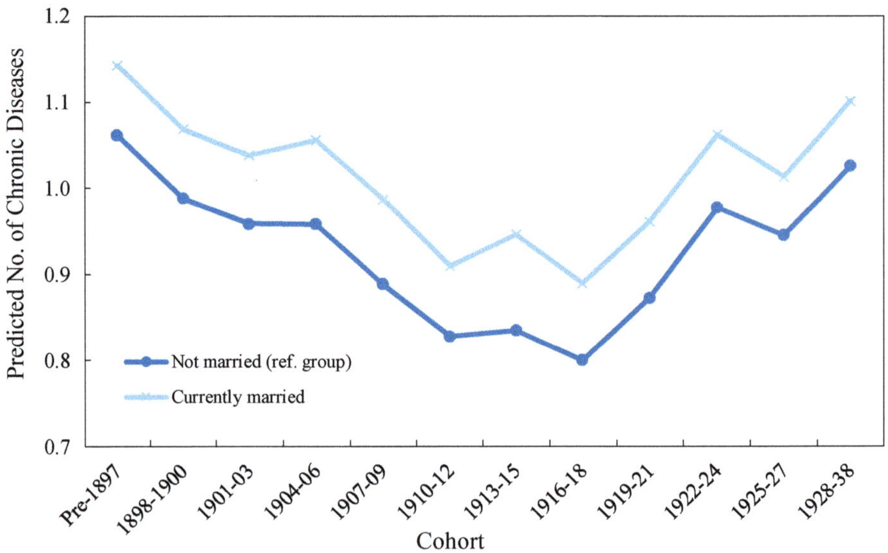

Fig. 5.15b Predicted number of chronic diseases by marital status from CCREMs over birth cohort. (Sources: 1998–2018 Chinese Longitudinal Healthy Longevity Survey (CLHLS))

subgroups, the 1916–18 birth cohort demonstrates a fewer number of chronic diseases than other birth cohorts.

A clear health advantage of married oldest old is shown when it comes to predicted cognitive function score depicted in Figs. 5.16a and 5.16b. As the figures show, the predicted cognitive function score is higher for married oldest old across all time periods and birth cohorts. Regarding the period effects, the predicted cognitive function score first shows a declining trend and an increasing pattern is observed in later years, suggesting that Chinese oldest old's cognitive function is getting better over time. As to the cohort effects, a bell curve is shown with the 1901–03 birth cohort showing the best cognitive function as compared to the other birth cohorts.

When predicting positive feeling scores, only period shows significant effects, which is depicted by Fig. 5.17. The cohort effects are not significant. As Fig. 5.17 shows, for most years, married oldest old show better positive feeling scores than non-married ones. Crossovers occur in years 2005, 2008 and 2011. Being consistent with previous figures, the predicted positive feeling scores also demonstrate a sharp decline in year 2008 due to an economic hardship as well as natural disasters occurred in that particular year. These results indicate that negative social events have significant impacts on all individuals, which may largely diminish the advantages of in positive feeling scores among married individuals.

Both significant period and cohort effects are shown when predicting negative feeling scores. Figures 5.18a and 5.18b shows significant period effects and cohort effects, respectively. According to the figure, non-married oldest old tend to have

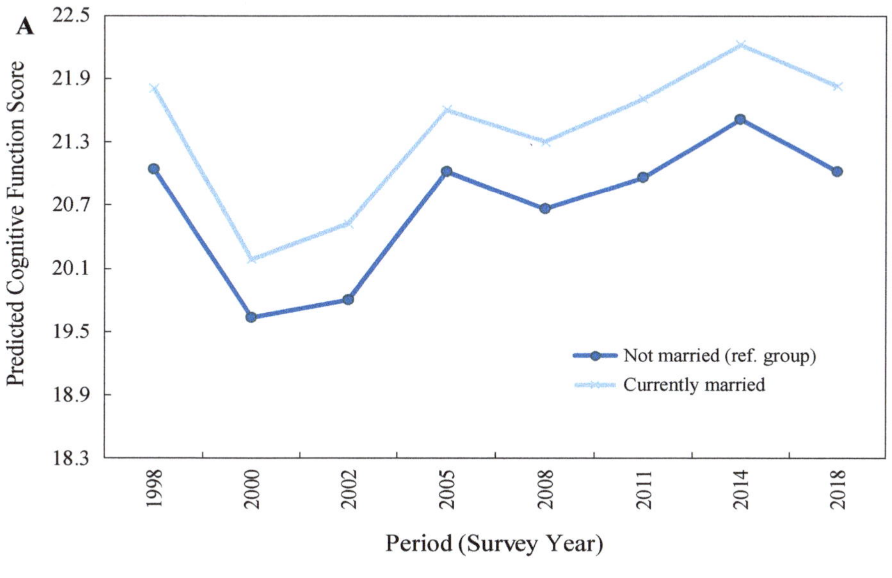

Fig. 5.16a Predicted cognitive function score by marital status from CCREMs over time period. (Sources: 1998–2018 Chinese Longitudinal Healthy Longevity Survey (CLHLS))

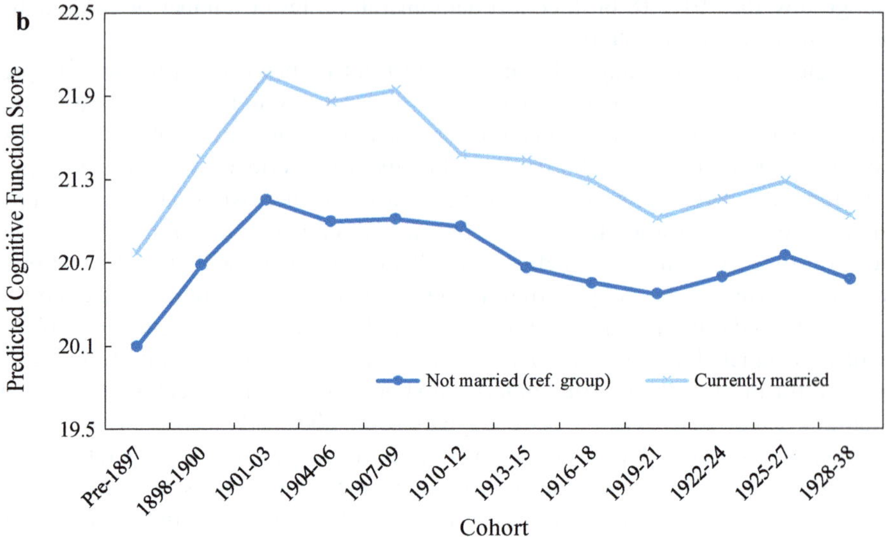

Fig. 5.16b Predicted cognitive function score by marital status from CCREMs over birth cohort. (Sources: 1998–2018 Chinese Longitudinal Healthy Longevity Survey (CLHLS))

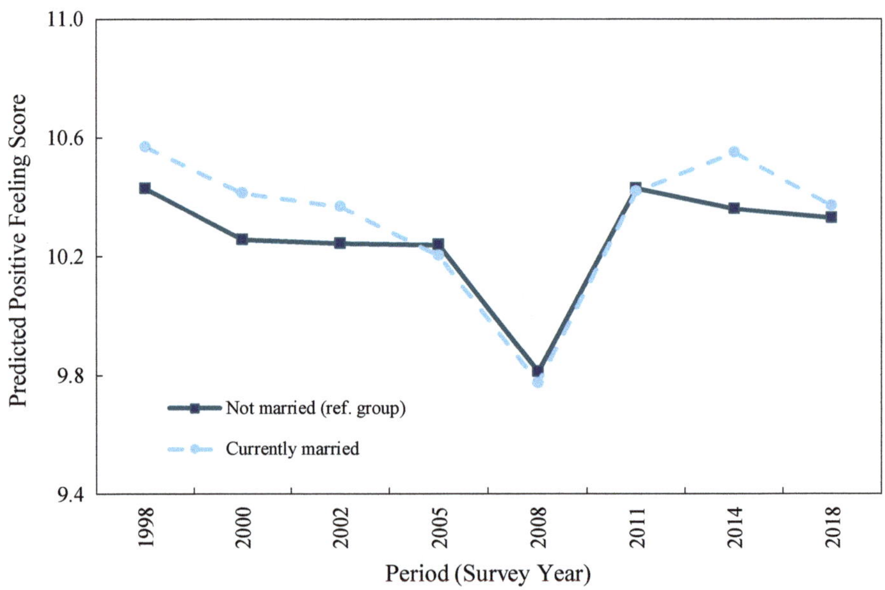

Fig. 5.17 Predicted positive feeling score by marital status from CCREMs over time period. (Sources: 1998–2018 Chinese Longitudinal Healthy Longevity Survey (CLHLS))

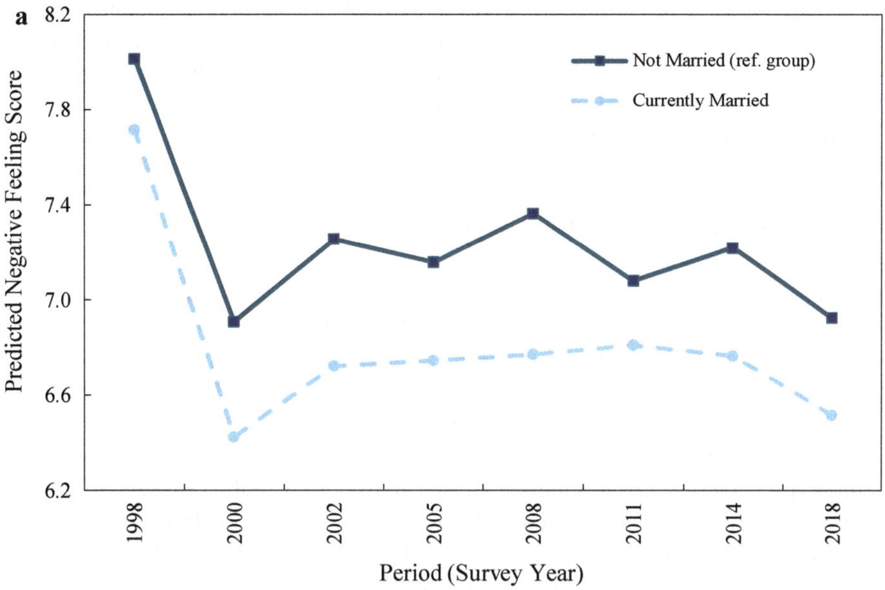

Fig. 5.18a Predicted negative feeling score by marital status from CCREMs over time period. (Sources: 1998–2018 Chinese Longitudinal Healthy Longevity Survey (CLHLS))

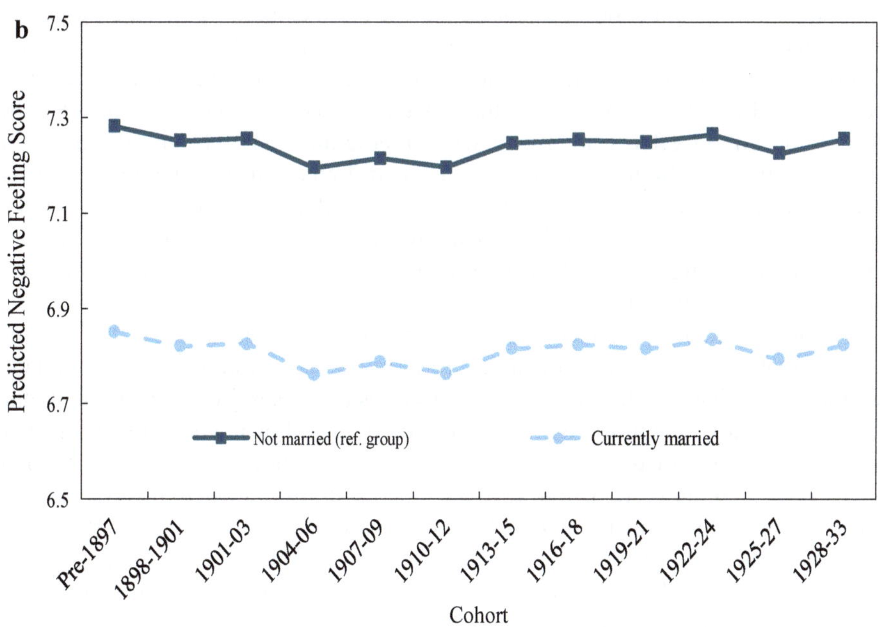

Fig. 5.18b Predicted negative feeling score by marital status from CCREMs over birth cohort. (Sources: 1998–2018 Chinese Longitudinal Healthy Longevity Survey (CLHLS))

higher predicted negative feeling scores than married individuals. A general downward trend is observed here, suggesting that Chinese oldest old's subjective welling is getting better. As to the cohort effects, they are not as apparent as period effects. The gaps in predicted negative feeling scores for both subgroups remain across birth cohorts with some fluctuations.

5.4 Conclusion and Discussion

This chapter has explored the period and cohort variations when using demographic characteristics to predict Chinese oldest old's health outcomes. This is one of the first studies that explore the age, period and cohort effects simultaneously when linking demographic characteristics to the oldest old's health. Several important findings can be drawn from this research.

First, the results largely corroborate findings from cross-sectional analyses that demographic factors are influential on the oldest old's health. Males, urban residents and married individuals generally show better health condition than females, rural citizens and unmarried respondents. These results support urban advantages and gender disparities in the oldest old's health status that have long been existing in China.

In addition to the above results, the most important findings are about the period and cohort variations in the linkages between demographic factors and the oldest old's health. The results show that when *sex* is used to predict the oldest old's health, significant period and cohort effects are shown when estimating SRH, chronic diseases and cognitive function. Significant period effects are found when predicting positive feeling scores and significant cohort effects are observed when estimating negative feeling scores. In terms of the linkage between *residence* and health, both period and cohort effects are found when predicting SRH, chronic diseases and cognitive function. Significant period effects are found when evaluating the oldest old's subjective wellbeing. As to *marital status*, significant period and cohort effects are observed when predicting the oldest old's chronic diseases, cognitive function and negative feeling scores. Only period effects are revealed when predicting SRH and positive feeling scores.

When specific period and cohort effects are taken into consideration, the findings suggest that health outcomes of males, rural residents and unmarried respondents are more sensitive to cohort and period variations than their female, urban and married counterparts. The most significant period effect is shown when predicting positive feeling scores of the oldest old. There is a sharp decline in year 2008 when using each demographic variable to estimate the respondent's positive feeling scores. This is likely to be caused by the economic hardship and the *Wenchuan* earthquake that occurred in that particular year. It implies that unlike physical health, the oldest old's subjective wellbeing is more likely to be influenced by certain social events occurred in unique periods.

In sum, this chapter has attempted to investigate how demographic factors impact Chinese oldest old's health and how period and cohort effects have played a role in shaping the respondents' health status. Although the research demonstrates period and cohort effects, the chapter has not explored why these effects exist when estimating the oldest old's health. Future research may investigate why such cohort and period effects have occurred to improve this current research. Even with these limitations, the current research contributes to existing studies by applying a longitudinal scope to explore how age, period and cohort have simultaneously impacted health of Chinese oldest old, controlling for important demographic and socioeconomic covariates. These findings help to advance findings and theories of the oldest old's health and possibly health of older adults in other social contexts.

References

Anderson, T. J., Saman, D. M., Lipsky, M. S., & Lutfiyya, M. N. (2015). A cross-sectional study on health differences between rural and non-rural U.S. counties using the county health rankings. *BMC Health Services Research, 15*, 441. https://doi.org/10.1186/s12913-015-1053-3

Bai, X. J., & Cheng, Y. S. (2019). Urban-rural regional differences in medical resource allocation and individual health of middle-aged and elderly people. *Journal of Xi'an Jiaotong University: Social Sciences Edition (in Chinese), 39*(2), 10.

Chen, P., Li, Z. Y., Jia, H., Sun, S. Y., & Li, A. L. (2020). Study on self-rated health status among elderly individuals aged 60 and above in impoverished rural areas of Xuyong County, Luzhou City. *Journal of Southwest Medical University(in Chinese), 1*, 36–39.

Cho, S., & Cho, O. H. (2022). Depression and quality of life in older adults with pneumoconiosis: The mediating role of death anxiety. *Geriatric Nursing, 44*, 215–220. https://doi.org/10.1016/j.gerinurse.2022.02.018

Fang, J. T., Li, W. X., Liu, X., & Shi, Z. M. (2015). Anxiety and depression among the elderly in Haidian District, Beijing. *Chinese Journal of Health Psychology (in Chinese), 23*(3), 447–450.

Fogelholm, M., Valve, R., Absetz, P., Heinonen, H., Uutela, A., Patja, K., Karisto, A., Konttinen, R., Makela, T., Nissinen, A., Jallinoja, P., Nummela, O., & Talja, M. (2006). Rural-urban differences in health and health behaviour: A baseline description of a community health-promotion programme for the elderly. *Scandinavian Journal of Public Health, 34*(6), 632–640. https://doi.org/10.1080/14034940600616039

Geldsetzer, P., Vaikath, M., Wagner, R., Rohr, J. K., Montana, L., Gomez-Olive, F. X., Rosenberg, M. S., Manne-Goehler, J., Mateen, F. J., Payne, C. F., Kahn, K., Tollman, S. M., Salomon, J. A., Gaziano, T. A., Barnighausen, T., & Berkman, L. F. (2019). Depressive symptoms and their relation to age and chronic diseases among middle-aged and older adults in rural South Africa. *The Journals of Gerontology. Series A, Biological Sciences and Medical Sciences, 74*(6), 957–963. https://doi.org/10.1093/gerona/gly145

Ho, S.-H. (2018). Correlations among self-rated health, chronic disease, and healthcare utilization in widowed older adults in Taiwan. *Journal of Nursing Research, 26*(5), 308–315. https://doi.org/10.1097/jnr.0000000000000248

Hossain, B., & James, K. S. (2022). Association between poor self-reported health and unmarried status among adults: Examining the hypothesis of marriage protection and marriage selection in the Indian context. *BMC Public Health, 22*(1), 1797. https://doi.org/10.1186/s12889-022-14170-0

Kohler, I. V., Payne, C. F., Bandawe, C., & Kohler, H. P. (2017). The Demography of mental health among mature adults in a low-income, High-HIV-Prevalence Context. *Demography, 54*(4), 1529–1558. https://doi.org/10.1007/s13524-017-0596-9

Krochalk, P. C., Li, Y., & Chi, I. (2008). Widowhood and self-rated health among Chinese elders: The effect of economic condition. *Australasian Journal on Ageing, 27*(1), 26–32. https://doi.org/10.1111/j.1741-6612.2007.00269.x

Li, C., Li, X., Quan, J., & Li, C. Y. (2013). Gender differences in health status of the left-behind elderly in rural areas. *Chinese Journal of Gerontology (In Chinese), 11*, 2619–2621.

Lv, X., Li, W., Ma, Y., Chen, H., Zeng, Y., Yu, X., Hofman, A., & Wang, H. (2019). Cognitive decline and mortality among community-dwelling Chinese older people. *BMC Medicine, 17*, 63.

Nakanishi, M., Ogawa, A., Sakai, M., Yoshii, H., Miyashita, M., Yamasaki, S., & Nishida, A. (2023). Depression and anxiety in older adults with dementia during the COVID-19 pandemic. *J Alzheimers Dis Rep, 7*(1), 307–315. https://doi.org/10.3233/ADR-230019

Oftedal, S., Kolt, G. S., Holliday, E. G., Stamatakis, E., Vandelanotte, C., Brown, W. J., & Duncan, M. J. (2019). Associations of health-behavior patterns, mental health and self-rated health. *Preventive Medicine, 118*, 295–303.

Perkins, J. M., Lee, H. Y., James, K. S., Oh, J., Krishna, A., Heo, J., Lee, J. K., & Subramanian, S. V. (2016). Marital status, widowhood duration, gender and health outcomes: A cross-sectional study among older adults in India. *BMC Public Health, 16*(1), 1032. https://doi.org/10.1186/s12889-016-3682-9

Saha, A., Rahaman, M., Mandal, B., Biswas, S., & Govil, D. (2022). Rural urban differences in self-rated health among older adults: Examining the role of marital status and living arrangements. *BMC Public Health, 22*(1). https://doi.org/10.1186/s12889-022-14569-9

Smith, K., & Victor, C. (2019). Typologies of loneliness, living alone and social isolation, and their associations with physical and mental health. *Aging & Society, 39*, 1709–1730.

Sun, Q., & Zhang, J. G. (2013). Dietary analysis and suggestions for improving the immunity of the elderly. *Journal of Food Research and Development (in Chinese), 34*(21), 105–106. https://doi.org/10.3969/j.issn.1005-6521.2013.21.030

Tang, D., & Chen, Z. (2011). Gender differences and age-related changes in the mental health of the elderly in the same life background. *Chinese Journal of Gerontology (In Chinese), 1*, 118–121.

Wu, X., Chao, J., Bao, M., Liu, Y., & Zhang, B. (2022). Study on gender difference and influencing factors of health status of the elderly in China. *Chinese Journal of Preventive Medicine, 1*, 13–19.

Xiong, Y. G., & Yang, X. (2016). Gender differences in health status of the urban elderly in China: An empirical analysis based on CHARLS data. *Journal of Jiangsu Institute of Administration, 4*, 56–65.

Xiu, D., Chow, A. Y. M., & Tang, S. (2020). Predictive factors for differential changes in grief symptoms following group bereavement intervention for Chinese widowed older adults. *Clinical Psychology & Psychotherapy, 27*(3), 267–277. https://doi.org/10.1002/cpp.2425

Yang, Y., & Land, K. C. (2006). A mixed models approach to the age-period-cohort analysis of repeatedc cross-section surveys, with an application to data on trends in verbal test scores. *Sociological Methodology, 36*(1), 75–97.

Yang, B., & Zhang, N. (2019). Gender differences in self-rated health of floating elderly: A multi-level model study based on health factors. *Population and Development (in Chinese), 2*, 20–30.

Yin, S. Y., Liang, Y. N., Ji, X. Q., Fan, W. Y., Zhang, F. L., & Mi, W. (2021). Health status and influencing factors of rural elderly individuals of different genders in Shandong Province. *Chinese Journal of Gerontology (In Chinese), 9*, 1965–1967. https://doi.org/10.3969/j.issn.1005-9202.2021.09.050

Yuan, H. (2021). Internet use and mental health problems among older people in Shanghai, China: The moderating roles of chronic diseases and household income. *Aging & Mental Health, 25*(4), 657–663.

Zeng, Y., Vaupel, J., Xiao, Z., Zhang, C., & Liu, Y. (2002). Sociodemographic and health profiles of the oldest old in China. *Population and Development Review, 28*, 251–273.

Zhang, J. H. (2018). Characteristics and comprehensive treatment of hypertension in the elderly. *Chinese Urban and Rural Enterprise Health (in Chinese), 33*(11). https://doi.org/10.16286/j.1003-5052.2018.11

Zhang, W., Li, S., & Feldman, M. (2005). Gender differences in activity of daily living of the elderly in rural China: Evidence from Chaohu. *Journal of Women & Aging, 17*(3), 73–89.

Zhang, J., Han, A., & Wu, X. (2021). Relation of depressive symptoms to intergenerational support and internet use in empty nest elderly. *Chinese Mental Health Journal, 35*(10), 838–843.

Zhang, P., Gao, S., Wang, H. J., & Su, M. Y. (2023). Gender differences and decomposition of self-rated health among the elderly in China. *Modern Preventive Medicine (in Chinese)*(10), 1848–1852+1902.

Zimmer, Z., & Kwong, J. (2004). Socioeconomic status and health among older adults in rural and urban China. *Journal of Aging and Health, 16*(1), 44–70.

References

Socioeconomic Status and the Oldest Old's Health

6

Prior studies have shown strong correlations between socioeconomic status (SES) and older adults' health. This chapter first reviews prior literature on the association between socioeconomic factors and older adults' health. It then conducts empirical analyses and explores when demographic characteristics are controlled, how period and cohort effects play a role in the relationship between SES and Chinese oldest old's health.

6.1 Previous Literature on Socioeconomic Characteristics and Older Adults' Health

Socioeconomic characteristics have been documented as important determinants of older adults' health. In previous literature, educational attainments and income level were commonly used indicators to measure one's socioeconomic status (SES). With regard to the correlation between educational attainments and older adults' health, multiple studies have demonstrated a significantly positive association between education and older adults' health outcomes in more developed countries (Harber-Aschan et al., 2020; Lima-Costa et al., 2012; Shankar et al., 2010). Older adults with higher education have been found to have a better immune function, which protects them from multiple types of diseases (Klopack et al., 2022). Individuals with higher education were also found to have more health related knowledge which helps them to maintain healthier lifestyles. Thus, individuals with higher education demonstrated better health outcomes (Meyer et al., 2014; Pampel et al., 2010; Tur-Sinai & Becker, 2022; Wroblewska, 2002). Prior research has also found a strong association between older adults' educational levels and their quality of life and subjective wellbeing. For example, education was estimated to account for approximately 30.0% of variations in older adults' subjective wellbeing (Brunello et al., 2016). Studies of Asian samples have also confirmed that higher levels of education

L. Zhang, *Living Longer and Healthier at Older Ages*, Advances in Studies of Aging and Health 3, https://doi.org/10.1007/978-3-031-69773-9_6

were associated with better SRH and lower mortality rates, although the associations between education and functional disabilities and chronic diseases were less pronounced (Hurt et al., 2004; Liang et al., 2000; Zimmer et al., 2004). A piece studying Mexican sample suggested that unlike more developed countries, there was a positive association between education and the risk of developing chronic diseases among Mexican older adults (Smith & Goldman, 2007).

In addition to education, income has also been considered as a major factor influencing older adults' health (Brasher et al., 2017). Researchers contended that older adults with higher income have the ability to purchase healthy food, which ensures them to obtain a balanced nutrition to prevent chronic diseases (Brinda et al., 2016). Higher-income older adults have been found to have a better access to healthcare resources, which offers them more opportunities for regular check-ups and treatments. Individuals with higher income often tended to have stronger self-health management beliefs that help them to maintain fairly good physical health conditions (Chen et al., 2015; Marks et al., 2019). Meanwhile, when older adults were more financially independent, they were found to be less likely to experience economic pressure and anxiety, which assisted them to maintain a positive mindset and good mental health status (Li & Bai, 2022).

Cumulative advantage theory has been applied to explain the health disparities due to socioeconomic reasons among individuals. Prior research showed that childhood experience correlated with individual health in older ages. Childhood is a critical period for formation of physiological functions, which is more susceptible to the influence of external environment (Perry, 2002). Health problems during childhood have been documented to increase the risk of having cardiovascular diseases, osteoarthritis and rheumatic chronic diseases among older adults (Liu & Li, 2022). Studies have also suggested an association between adverse childhood experience and depression in older ages. Adverse experiences in childhood resulted in a lack of security and trust in others, which gradually caused self-isolation and an increased risk of developing depression (Xiang & Wang, 2021). Early experiences of economic difficulties were observed to link to an increased risk of depression in later years (Liu & Li, 2021). This is also referred as the "long arm of childhood" (Hayward & Gorman, 2004). Since most existing studies are cross-sectional, little is known about the period and cohort effects in the correlation between SES and older adults' health. This chapter aims to improve the existing studies by conducting APC analyses to investigate the period and cohort effects.

6.2 Data, Measures and Methods

6.2.1 Data

The analyses of this chapter are carried out by analyzing data from eight (1998, 2000, 2002, 2005, 2008, 2011–12, 2014 and 2018) waves of the Chinese Longitudinal Healthy Longevity Survey (CLHLS). The CLHLS was undertaken in 631 randomly selected counties and cities of 22 provinces and municipalities directly under the

Central Government in China. These 22 provinces and municipalities include Liaoning, Jilin, Heilongjiang, Hebei, Beijing, Tianjin, Shanxi, Shaanxi, Shanghai, Jiangsu, Zhejiang, Anhui, Fujian, Jiangxi, Shandong, Henan, Hubei, Hunan, Guangdong, Guangxi, Sichuan, and Chongqing. The survey covers roughly half the counties and cities of those provinces, and the sample areas represent 85.0% of the total population of China.

The1998 baseline survey includes 9,073 oldest old individuals aged 80 and older. In 2000, 4,844 of those elders were reinterviewed, and 6,372 elders were newly added interviewees. For more detailed description of the data, see Zeng et al. (2002). This chapter focuses on studying oldest old aged 85 to 105 since prior literature considered persons who reported age 106 or higher as invalid cases (Zeng et al., 2002).

6.2.2 Measures

As to measures of socioeconomic status, the study uses the highest degree received to represent the respondent's *educational attainment*. In this dataset, respondents are categorized into three groups according to their educational attainments, which are: respondents with elementary school degrees or less, respondents with middle school degrees, and respondents who received high school degrees or above. Those who had elementary school degrees or less are treated as the reference group. I*ncome* is a commonly used measure of SES, which is usually measured in multiple ways. For instance, income can be measured by individual or family income. Some people also used the logged form of income to avoid a skewed distribution of income for a population. In this study, however, family income information is only available in recent five waves of CLHLS. Such information is not asked in the first three waves of the survey. Since the study is interested in examining the period effect, spanning the eight waves, the income variable has to be dropped in this analysis.

Employment or *occupation* is also commonly used to capture the respondent's SES. In the CLHLS questionnaire, the respondent was asked about his/her occupation before retirement. There is also a question asking occupation of the respondent's father. These measures can be used as covariates representing the respondent's SES. But after conducting preliminary analyses, I found the occupation variable is highly correlated with the education variable. Thus, the occupation variable is not used in this research.

There is a question in CLHLS survey asking the respondent if he or she *often went to bed hungry in childhood*. Prior literature used this measure to represent the respondent's SES and found that going to bed hungry in childhood was significantly linked to worse health in older ages (Zhang, 2021). Thus, this measure is used as an indicator of the oldest old's SES in this study.

Health outcomes are measured by five indictors, which are consistent with measures used in previous chapters, including self-rated health (SRH), number of chronic diseases, cognitive function, and positive and negative feelings which are considered are measures of subjective wellbeing (Lv et al., 2019; Oftedal et al., 2019;

Smith & Victor, 2019; Yuan, 2021). Similar to previous chapters, the respondent's *self-rated health* is coded as a continuous variable (1 = very bad, 5 = very good). *Cognitive function* of the respondent is measured by using the Chinese version of the Mini-Mental State Examination (MMSE). The MMSE is adapted from Folstein, Folstein, and McHugh (Folstein et al., 1975) and tests four aspects of cognitive functioning: orientation, calculation, recall, and language. The total possible score on the MMSE is 30, with lower scores indicating poor cognitive ability. Responses of "unable to answer" are coded as incorrect answers. Chronic disease is measured by number of chronic diseases reported by the respondent, which ranges from 0 to 13. Positive and negative feeling scores are also measured in the same way that was presented in the previous chapter.

6.2.3 Methods

A cross-classified random-effects model (CCREM) is used to take into account the embeddedness of respondents in the CLHLS surveys within a time period (Yang & Land, 2006). Specifically, the model estimates fixed effects of individual-level covariates and random effects of period (or time). Individuals are the level-1 unit of analysis. Periods (or time) and birth cohorts become the level-2 units of analysis. In the analysis of the CLHLS data, an individual respondent is nested in and cross-classified by a higher level context, which is period (or time) and birth cohort.

Beyond providing random period effects, CCREMs allow for random effects of independent variables, which is used to test for cross-period variations in the effects of key independent variables, i.e. socioeconomic (SES) variables. Period-specific random-effects coefficients and cohort–specific random-effects coefficients suggest potential across-period and across-cohort changes in the effects of SES factors on the oldest old's health status, respectively. This study specifies models of the oldest old's health status as a quadratic function of age.

Several models are presented in the analysis. The first model only includes age and age-squared as fixed effects. This model measures the overall effects of period on the oldest old's health status, controlling for age and age squared variables. The next models include SES measures, controlling for the effects of age, age squared term and demographic variables. The full individual level or level-1 equation can be expressed as:

$$Y_{ijk} = \beta_{0jk} + \beta_{1jk}AGE + \beta_{2jk}AGE^2 + \beta_{3jk}X + \sum_{p=10}^{p} \beta_p Xp + e_{ijk} \tag{6.1}$$

where Y_{ijk} represents health status of the *i*th respondent for i = 1, …, n*j* individuals within the *j*th period for j = 1, …. β_{0j} is the intercept or for respondents in period *j*; "AGE" and "AGE^2" symbolize age and age-squared, respectively. "X" indicates independent variables; i.e., SES variables. β_p represents other individual-level fixed effects where *k* is the maximum number of covariates. e_{ij} represents an individual level random error term.

The level-2 model can be expressed as:

$$\beta_{0j} = \gamma_0 + p_{0j} + c_{0k} \quad (6.2)$$

This model specifies the overall mean vary from period to period and from cohort to cohort. γ_0 is the model intercept, which is the expected mean at the zero value of all level-1 variables averaged over all periods and cohorts. Here p_{0j} and c_{0k} are the residual random effects of period and cohort, respectively. The cell mean, $\beta_{0j,}$ is equal to the sum of the overall mean or intercept (γ_0), and the residual random effect of period j (p_{0j}) and the residual random effect of cohort (c_{0k}). The residual random effects of period j (p_{0j}) and cohort k (c_{0k}) examine the effects of each period and cohort on a specific health measure. When adding (6.2) to (6.1), it yields the level-1 and level-2 combined model:

$$Y_{ijk} = \gamma_0 + \beta_{0jk} + \beta_{1jk}AGE + \beta_{2jk}AGE^2 + \beta_{3jk}X + p_{0j} + c_{0k} + e_{ijk} \quad (6.3)$$

where, within each survey year j, respondent i's health is modeled as a function of his or her age, age-squared, the SES variables and other control variables. In the above model, γ_0 represents the fixed-effects coefficients for SES variables; the symbols of p_{0j} and c_{0k} represent the period-specific and cohort-specific effects of SES variables, respectively. The model tests for random effects of SES variables on the respondent's health across periods and cohorts. All continuous variables are sorted and centered around their overall mean.

Since there are eight waves of survey data, survey years for the eight waves are considered as the eight periods. I have developed 13 birth cohorts in the analyses (pre-1897, 1898–1900, 1901–03, 1904–06, 1907–09, 1910–12, 1913–15, 1916–18, 1919–21, 1922–24, 1925–27, 1928–30 and 1931–33). The 1934 and later birth cohorts are dropped from the analyses since those cohorts do not fall into the age range of 85 to 105. When showing descriptive results, all 13 birth cohorts are shown in graphs. When conducting regression analyses, the 1928–30 and 1931–33 birth cohorts are combined into one birth cohort considering the number of respondents who were born in and after 1928 is limited.

6.3 Results

6.3.1 Descriptive Statistics

6.3.1.1 Basic Statistics of Variables

Table 4.1 in Chap. 4 have presented descriptive statistics for all variables used in this research. In total, 58,485 oldest old have been surveyed in eight waves of CLHLS. Since CLHLS is a sampling survey and the study does not use weighted percentages, the percentage distributions refer to the sample. Based on descriptive results, the respondents' SES is low with the average schooling year being 1.4. The majority of the respondent reported elementary school education or less. And 67.9%

of them reported frequently going to bed hungry in childhood. Detailed descriptive information for period and cohort variations in SES variables will be depicted in graphs in the following sections.

6.3.1.2 Period and Cohort Variations in Socioeconomic Profiles of the Oldest Old

This section moves to graphic presentations of period and cohort variations in SES of the respondents. Figures 6.1a, 6.1b and 6.1c show educational attainments of the oldest old by period and across birth cohorts. As it is shown in these three figures, over 85.0% of the oldest old claimed their educational level as elementary school or less. Figure 6.1a shows that the percentage of respondents who had elementary school education or less has been declining slightly across birth cohorts. When period variations are taken into consideration, an increasing trend is observed in earlier waves. There is a clear decreasing trend since the 2014 wave.

Figure 6.1b shows percentages of the oldest old who had middle school education for various birth cohorts by time period. For most birth cohorts, a declining trend is observed and an increasing trend is found since year 2014. For this particular group, cohort variations are more noticeable than period variations. Generally speaking, younger birth cohorts tend to report higher percentages of middle school educational attainments than older birth cohorts.

Period and cohort variations are displayed in Fig. 6.1c for the group having high school or above education. The percentage shares of this group range from 1.0% to 9.5% of all respondents during the eight studied waves. Similar to the subgroup with

a

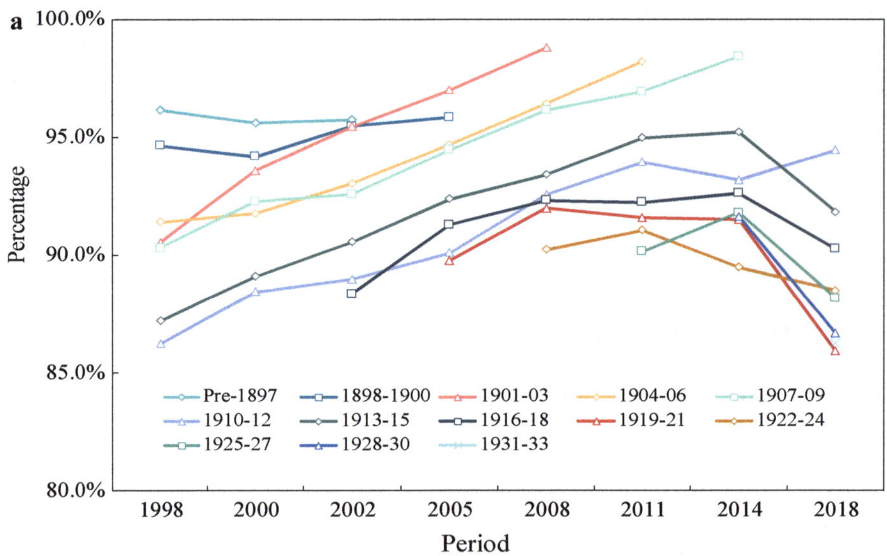

Fig. 6.1a Percentage of respondents with elementary school education or less by time period and across birth cohorts (Unweighted). (Sources: 1998–2018 Chinese Longitudinal Healthy Longevity Survey (CLHLS))

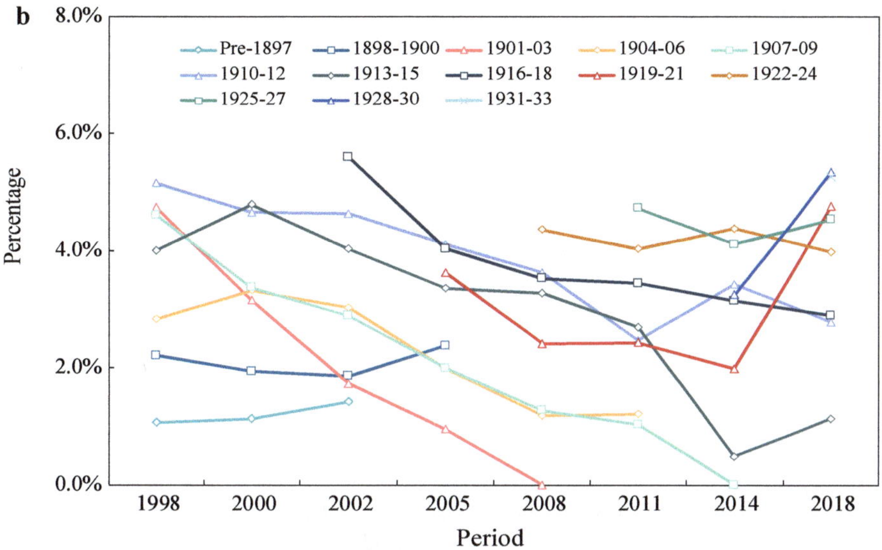

Fig. 6.1b Percentage of respondents with middle school education by time period and across birth cohorts. (Sources: 1998–2018 Chinese Longitudinal Healthy Longevity Survey (CLHLS))

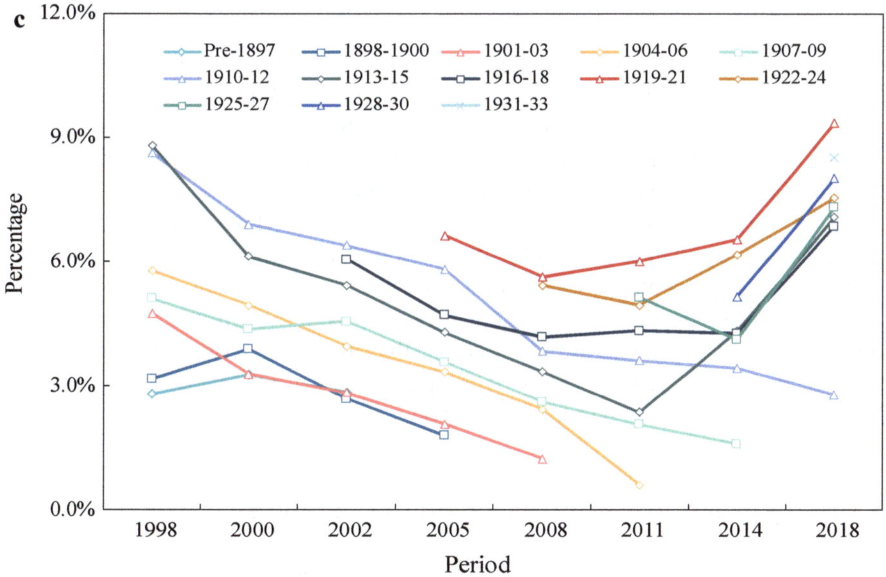

Fig. 6.1c Percentage of respondents with high school education or above by time period and across birth cohorts (Unweighted). (Sources: 1998–2018 Chinese Longitudinal Healthy Longevity Survey (CLHLS))

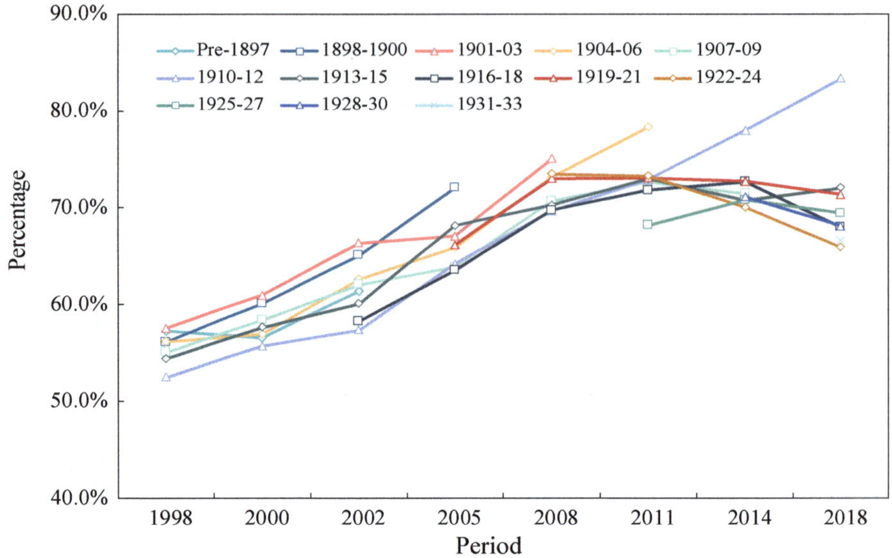

Fig. 6.2 Percentage of respondents who frequently went to bed hungry by time period and across birth cohorts (Unweighted). (Sources: 1998–2018 Chinese Longitudinal Healthy Longevity Survey (CLHLS))

middle school education, clear cohort variations are shown with younger birth cohorts reporting higher percentage shares of having high school education. A downward trend is shown for the first five waves. Since year 2011, an upward trend is observed, suggesting that the percentage shares of respondents with high school or college education has been increasing over time.

Figure 6.2 depicts the percentage shares of the oldest old who reported frequently going to bed hungry in childhood by time period and across birth cohorts. The figure shows that the percentages of respondents who reported frequently going to bed hungry in childhood range from 52.4% to 83.3% from 1998 to 2018. For this variable, the cohort variations are less evident than the period variations. Over time, there is an upward trend in terms of the percentage of respondents who reported frequently going to bed hungry in childhood.

6.3.2 Regression Results Based on APC Analyses

This section of the analyses puts each of the SES variables into the second level of the model to investigate the period and cohort effects. Tables 6.1 and 6.2 show regression results from the CCREMs when including education and frequently going to bed hungry in childhood variables in the second level of the model, respectively. Only significant period and cohort variations are presented by graphs.

Table 6.1 Regression results of health outcome measures from CCREMs when education is included in level 2: Chinese Aged 85–105

	Self-rated health		Chronic disease		Cognitive function		Positive feelings		Negative feelings	
Variables	Coef.	P Value	Coef.	P Value	Coef.	P Value	Coef.	P Value	Coef.	P Value
Fixed effect component										
Intercept	3.413	<0.001	0.923	<0.001	20.710	<0.001	10.262	<0.001	7.240	<0.001
Control variables										
Age	0.002	0.060	−0.015	<0.001	−0.355	<0.001	−0.018	<0.001	−0.004	0.071
Age ^{2}	0.000	0.595	0.000	0.149	0.001	0.489	0.000	0.619	0.000	0.258
Sex (male = 1)	0.089	<0.001	−0.025	0.040	2.110	<0.001	0.221	<0.001	−0.319	<0.001
Residence (urban =1)	0.054	<0.001	0.301	<0.001	0.545	<0.001	0.384	<0.001	−0.319	<0.001
Being married	−0.033	0.008	0.079	<0.001	0.667	<0.001	0.061	0.065	−0.444	<0.001
Education (ref. = primary school or below)										
Middle school	0.075	0.057	0.304	0.030	1.875	<0.001	0.500	<0.001	−0.262	0.017
Higher education	0.042	0.111	0.477	0.001	1.655	<0.001	0.456	<0.001	−0.385	<0.001
Often going to bed hungry as a child (ref. = yes)	−0.059	<0.001	−0.041	<0.001	−0.791	<0.001	−0.225	<0.001	0.242	<0.001

(continued)

Table 6.1 (continued)

	Self-rated health		Chronic disease		Cognitive function		Positive feelings		Negative feelings	
Variables	Coef.	P Value	Coef.	P Value	Coef.	P Value	Coef.	P Value	Coef.	P Value
Random effect component										
Period										
Intercept	0.008	0.034	0.058	0.033	0.434	0.037	0.043	0.034	0.124	0.032
Education										
Middle school	0	–	0.083	0.052	0	–	0.008	0.336	0	–
Higher education	0.001	0.324	0.062	0.041	0.147	0.181	0	–	0.011	0.268
Cohort										
Intercept	0.000	0.101	0.007	0.020	0.108	0.034	0.003	0.105	0.001	0.160
Education level										
Middle school	0.006	0.142	0.011	0.176	0	–	0	–	0.038	0.149
Higher education	0	–	0	–	0	–	0.040	0.100	0.000	0.486
AIC	128,757.2		174,692.4		316,199.6		223,085.8		210,168.9	
BIC	128,747.2		174,680.4		316,191.6		223,075.8		210,156.9	

Source: 1998–2018 Chinese longitudinal healthy longevity survey (CLHLS)

Table 6.2 Regression results of health outcome measures from CCREMs when frequently going to bed hungry in childhood is included in level 2: Chinese aged 85–105

Variables	Self-rated health		Chronic disease		Cognitive function		Positive feelings		Negative feelings	
	Coef.	P Value	Coef.	P Value	Coef.	P Value	Coef.	P Value	Coef.	P Value
Fixed effect component										
Intercept	3.418	<0.001	0.964	<0.001	20.721	<0.001	10.264	<0.001	7.241	<0.001
Control variables										
Age	0.001	0.093	−0.015	<0.001	−0.357	<0.001	−0.017	<0.001	−0.004	0.072
Age ^{2}	0.000	0.565	0.000	0.110	0.001	0.481	0.000	0.645	0.000	0.253
Sex (male = 1)	0.088	<0.001	−0.026	0.034	2.111	<0.001	0.221	<0.001	−0.318	<0.001
Residence (urban =1)	0.054	<0.001	0.300	<0.001	0.546	<0.001	0.385	<0.001	−0.319	<0.001
Being married	−0.033	0.008	0.081	<0.001	0.667	<0.001	0.061	0.062	−0.444	<0.001
Education (ref. = primary school or below)										
Middle school	0.079	<0.001	0.279	<0.001	1.867	<0.001	0.499	<0.001	−0.285	<0.001
High school	0.039	0.050	0.476	<0.001	1.651	<0.001	0.482	<0.001	−0.389	<0.001

(continued)

Table 6.2 (continued)

Variables	Self-rated health		Chronic disease		Cognitive function		Positive feelings		Negative feelings	
	Coef.	P Value	Coef.	P Value	Coef.	P Value	Coef.	P Value	Coef.	P Value
Often going to bed hungry as a child (ref. = yes)	−0.066	0.001	−0.080	0.187	−0.800	<0.001	−0.233	<0.001	0.242	<0.001
Random effect component										
Period										
Intercept	0.007	0.034	0.116	0.033	0.450	0.036	0.044	0.033	0.126	0.032
Often going to bed hungry as a child	0	–	0.022	0.041	0	–	0	–	0	–
Cohort										
Intercept	0.000	0.350	0.008	0.024	0.091	0.042	0.001	0.276	0.001	0.171
Often going to bed hungry as a child	0.001	0.082	0.001	0.208	0.030	0.159	0.004	0.144	0	–
AIC	128,752.9		174,723.0		316,199.5		223,086.7		210,166.4	
BIC	128,744.9		174,713.0		316,191.5		223,078.7		210,160.4	

Source: 1998–2018 Chinese Longitudinal Healthy Longevity Survey (CLHLS)

6.3.2.1 Period and Cohort Variations in Health Disparities by Educational Attainments

This section shows the period and cohort effects in the linkage between educational attainments and the oldest old's health outcomes, controlling for age, demographic factors and other SES variables. Similar to previous sections, only significant period and cohort effects are depicted in figures. Figures 6.3a and 6.3b show significant period and cohort effects when using educational attainments to predict SRH, respectively.

Unlike what I expected initially that individuals with the highest educational attainment have the best predicted SRH. The figure, however, indicates that the oldest old with middle school education have the highest predicted SRH, followed by those with high school education or above. The oldest old with primary school education or less show the lowest level of predicted SRH. Figure 6.3a also indicates that the health disparities among three subgroups remain with a general declining trend for predicted SRH among all three subgroups.

Figure 6.3b depicts an interesting story when it comes to the cohort effects. It seems that the predicted SRH for the oldest old with middle school education is more sensitive to cohort changes as compared to the other two subgroups. Three obvious declines occur among the 1907–09, 1922–24 and 1928–33 birth cohorts. It worth investigating why SRH of the oldest old with middle school education are more sensitive to cohort variations.

Significant period and cohort effects in predicted number of chronic diseases are shown in Figs. 6.4a and 6.4b, respectively. As the figures show, educational attainments are negatively associated with the predicted number of chronic diseases.

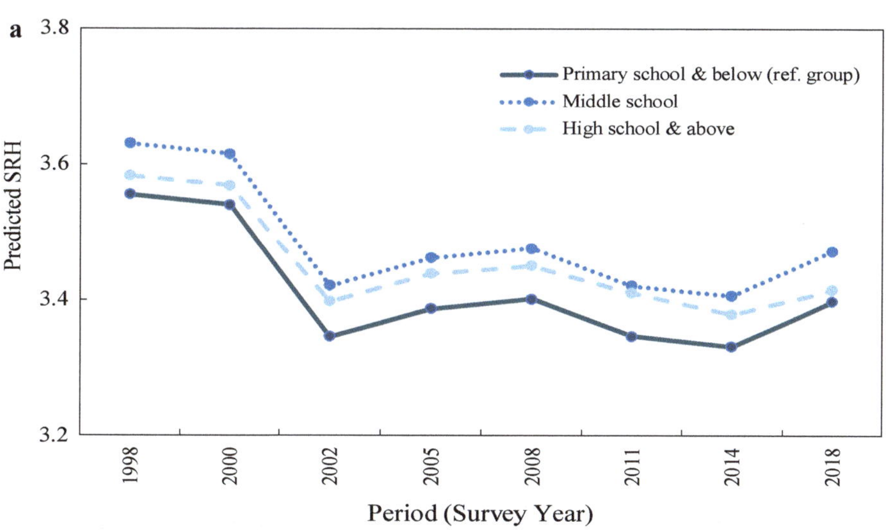

Fig. 6.3a Predicted self-rated health by educational attainment from CCREMs over time period. (Sources: 1998–2018 Chinese Longitudinal Healthy Longevity Survey (CLHLS))

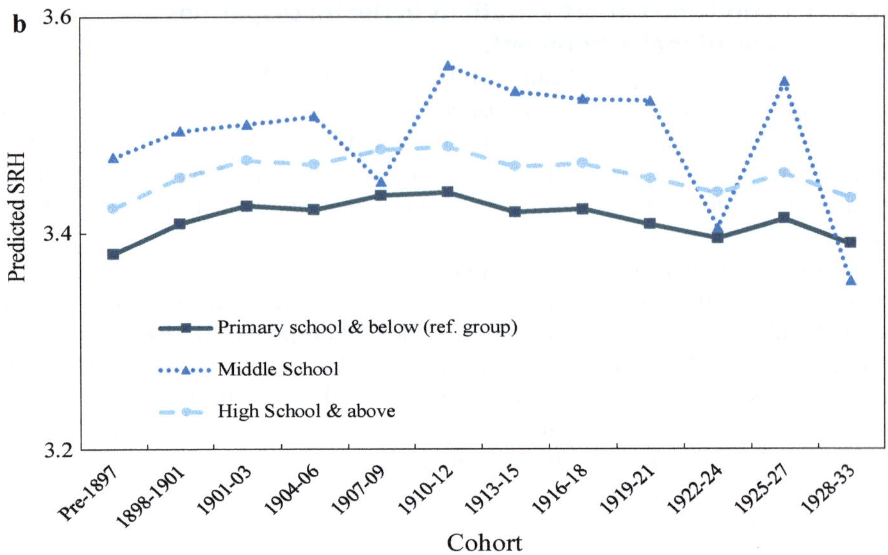

Fig. 6.3b Predicted self-rated health by educational attainment from CCREMs over birth cohort. (Sources: 1998–2018 Chinese Longitudinal Healthy Longevity Survey (CLHLS))

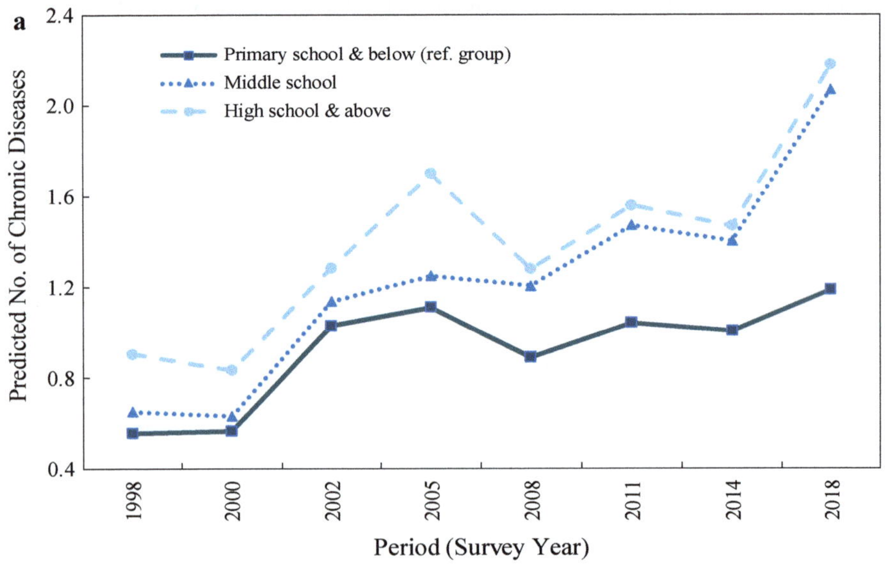

Fig. 6.4a Predicted number of chronic diseases by educational attainment from CCREMs over time period. (Sources: 1998–2018 Chinese Longitudinal Healthy Longevity Survey (CLHLS))

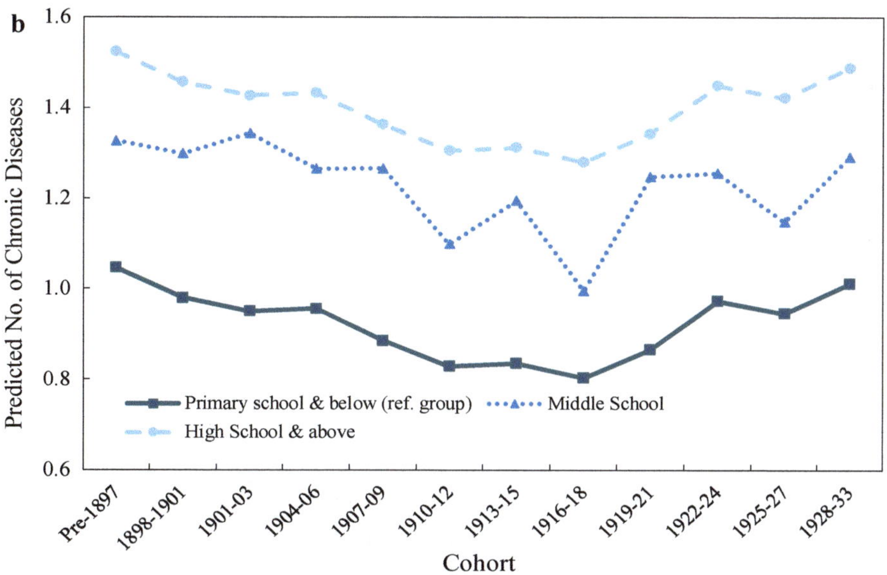

Fig. 6.4b Predicted number of chronic diseases by educational attainment from CCREMs over birth cohort. (Sources: 1998–2018 Chinese Longitudinal Healthy Longevity Survey (CLHLS))

This is contradictory to findings based on studying Western oldest old sample. This is probably because China is going through economic and social transitions. People with higher educational attainments tend to consume more high-fat and high-energy food, which has been considered as the privilege of people with higher SES (Chen et al., 2010). Consequently, they report more chronic diseases due to their diet patterns. The other reason could be that individuals with higher education are economically doing better, thus, diagnosis of chronic diseases is more affordable to them. As a result, the statistics show that they report a greater number of chronic diseases. The period effects are mainly shown by an increasing trend of in predicted number of chronic diseases for both subgroups with some fluctuations. The cohort effects are shown by a downward trend for earlier birth cohorts and an upward trend is observed for more recent birth cohorts. As compared to the other two subgroups, the predicted number of chronic diseases for the oldest old with middle school education is more sensitive to cohort variations.

As to predicted cognitive function scores, only period effects are found to be significant, which is depicted by Fig. 6.5. The figure shows that the oldest old with primary school or lower education have the lowest predicted cognitive function scores among all three subgroups. The disparities in cognitive function for the other two subgroups are small with crossovers. This finding suggests that having middle school education may be a threshold differentiating the studied oldest old's cognitive function status. Period wise, fluctuations are shown in predicted cognitive function scores with years 2005 and 2014 showing higher cognitive function scores for all three subgroups.

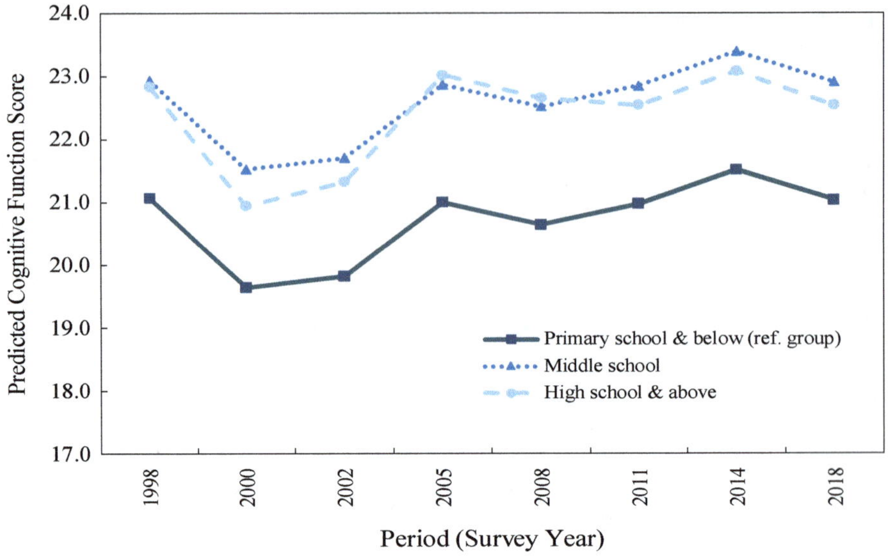

Fig. 6.5 Predicted cognitive function Score by educational attainment from CCREMs over time period. (Sources: 1998–2018 Chinese Longitudinal Healthy Longevity Survey (CLHLS))

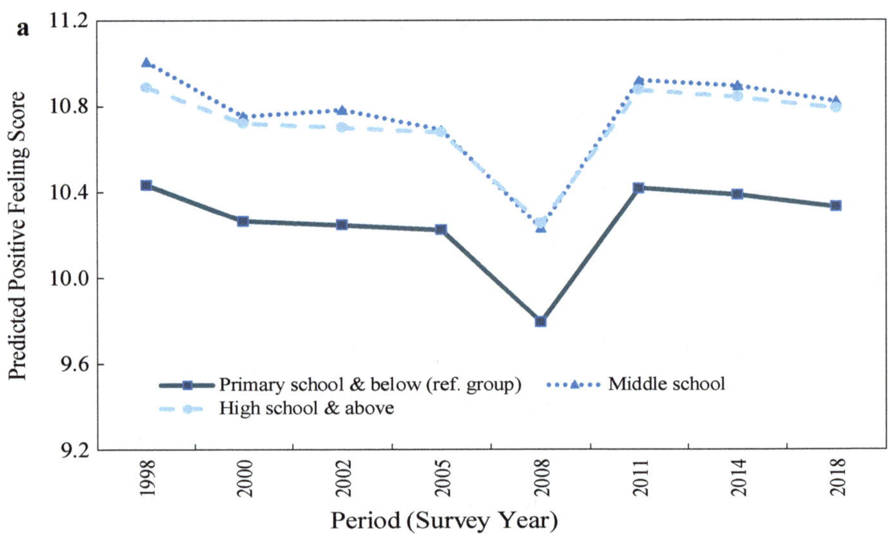

Fig. 6.6a Predicted positive feeling Score by educational attainment from CCREMs over time period. (Sources: 1998–2018 Chinese Longitudinal Healthy Longevity Survey (CLHLS))

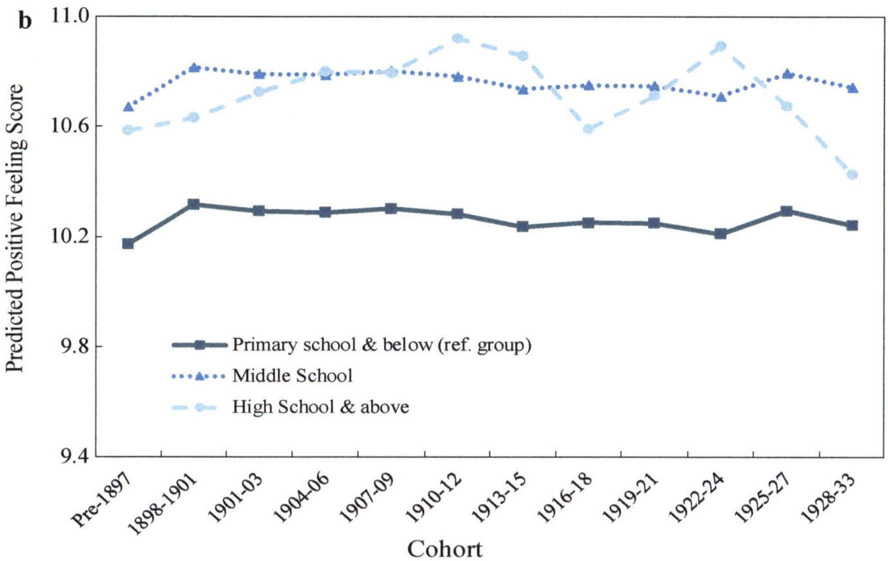

Fig. 6.6b Predicted positive feeling Score by educational attainment from CCREMs over birth cohort. (Sources: 1998–2018 Chinese Longitudinal Healthy Longevity Survey (CLHLS))

Figures 6.6a and 6.6b show significant period and cohort effects when predicting positive feeling scores of the respondents with different educational levels, controlling for age and other demographic and SES factors. The results show that the oldest old with primary school or lower education have the lowest positive feeling scores, indicating that the oldest old's subjective wellbeing is positively linked to one's educational attainment. The positive feeling scores for the respondents with middle school and high school education largely overlap with each other in Fig. 6.6a. There is an apparent dip in year 2008, which demonstrates significant period effects. As to the cohort effects, they are mainly shown among the oldest old with high school education. It suggests that the oldest old with high school or above educational attainments are more likely to be affected by cohort variations when their subjective wellbeing is taken into consideration.

The CCREM regression results using educational attainments to predict the oldest old's negative feeling scores, controlling for age and other demographic and SES covariates are shown in Figs. 6.7a and 6.7b. These two figures highlight significant period and cohort effects, respectively. Consistent to results predicting positive feeling scores, the oldest old with primary school or lower educational attainments have the highest negative feelings scores, followed by the ones with middle school and high school educational attainments. There is a considerable drop of predicted negative feeling scores from year 1998 to 2000. No clear increasing or decreasing trends are observed for the following years. In terms of the cohort effects,

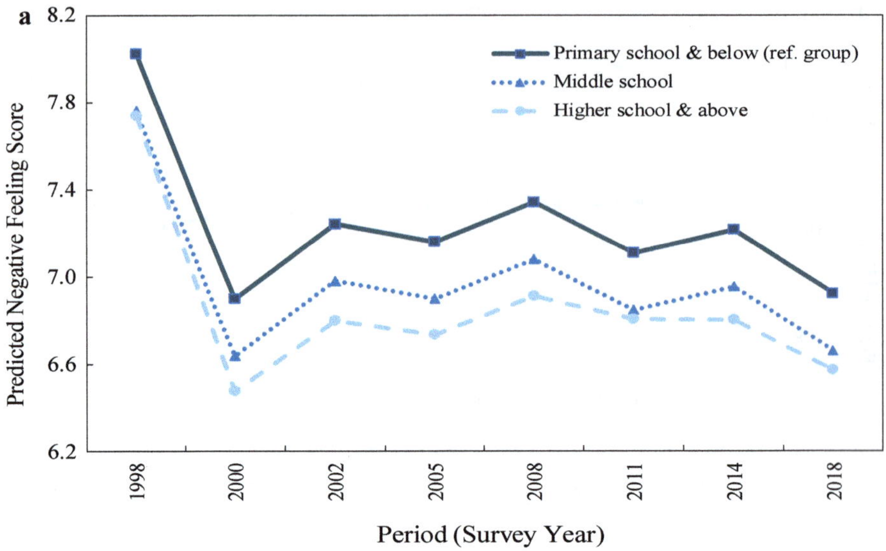

Fig. 6.7a Predicted negative feeling Score by educational attainment from CCREMs over Time Period. (Sources: 1998–2018 Chinese Longitudinal Healthy Longevity Survey (CLHLS))

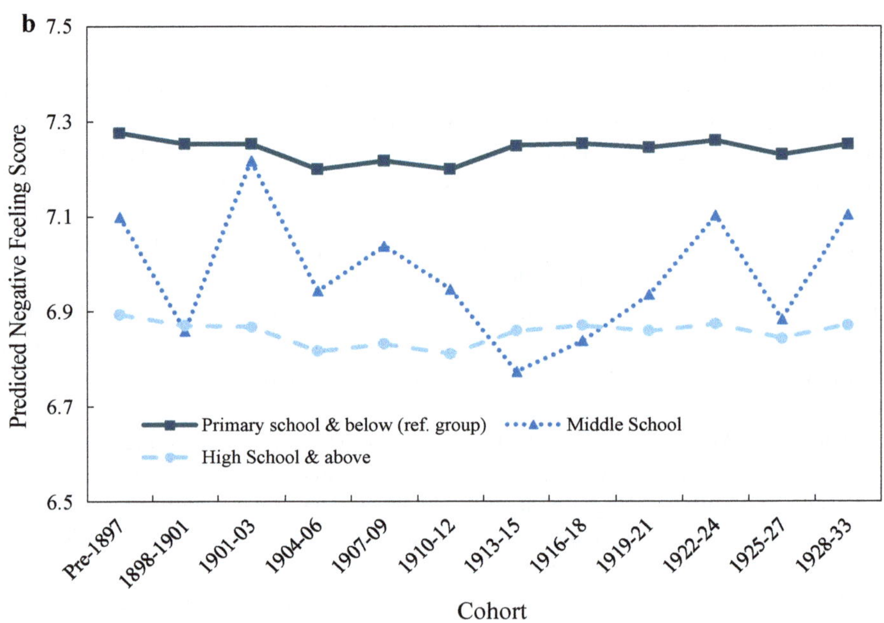

Fig. 6.7b Predicted negative feeling Score by educational attainment from CCREMs over birth cohort. (Sources: 1998–2018 Chinese Longitudinal Healthy Longevity Survey (CLHLS))

significant variations for negative feeling scores are mainly found among the subgroup with middle school education. The finding again suggests that the oldest old with middle school education are more sensitive to cohort changes than other subgroups.

6.3.2.2 Period and Cohort Variations in Health Disparities by Frequently Going to Bed Hungry in Childhood

This section studies the period and cohort effects in the link between the variable "frequently going to bed hungry in childhood" and the respondent's health outcomes, controlling for age, other demographic and the education variables. Similar to previous sections, only significant period and cohort effects are shown in figures. When predicting the oldest old's SRH, only cohort effects are found to be significant. As Fig. 6.8 indicates, the oldest old who frequently went to bed hungry show worse SRH as compared to those who seldom did so. The cohort effects are mainly shown among the subgroup that frequently went to bed hungry. The 1910–12 birth cohort has better predicted SRH than its earlier and later birth cohorts.

Regarding the predicted number of chronic diseases, both period and cohort effects are significant, which are shown in Figs. 6.9a and 6.9b, respectively. Figure 6.9a shows that before year 2002, those who frequently went to be hungry tend to have a greater predicted number of chronic diseases. After 2002, the reversed pattern seems to be true, i.e., those who were economically doing better tend to have a greater number of chronic diseases. According to Fig. 6.9b, the oldest old who seldom went to bed hungry as a child show a greater number of chronic diseases

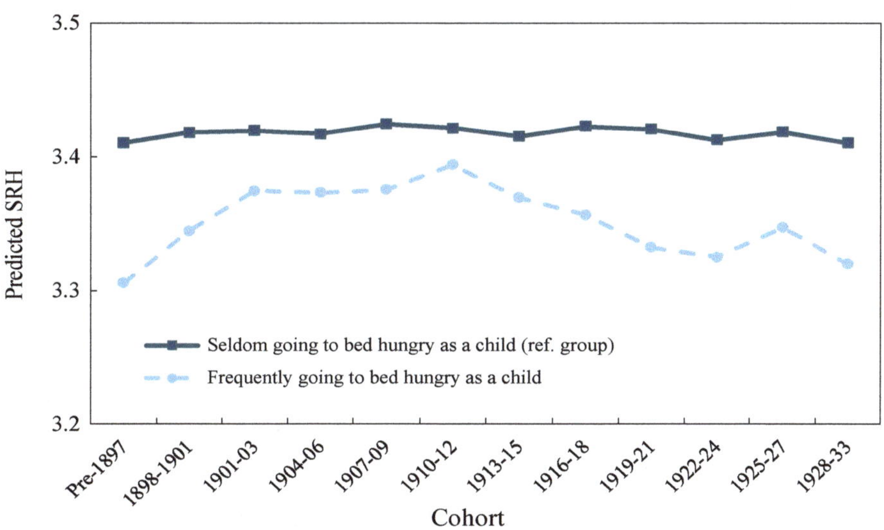

Fig. 6.8 Predicted SRH by frequently going to bed hungry in childhood from CCREMs over birth cohort. (Sources: 1998–2018 Chinese Longitudinal Healthy Longevity Survey (CLHLS))

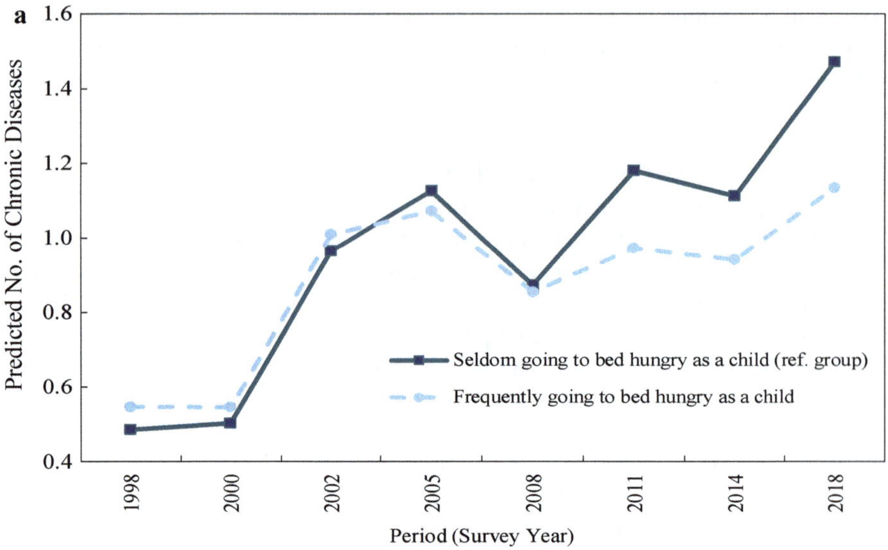

Fig. 6.9a Predicted number of chronic diseases by frequently going to bed hungry in childhood from CCREMs over Time Period. (Sources: 1998–2018 Chinese Longitudinal Healthy Longevity Survey (CLHLS))

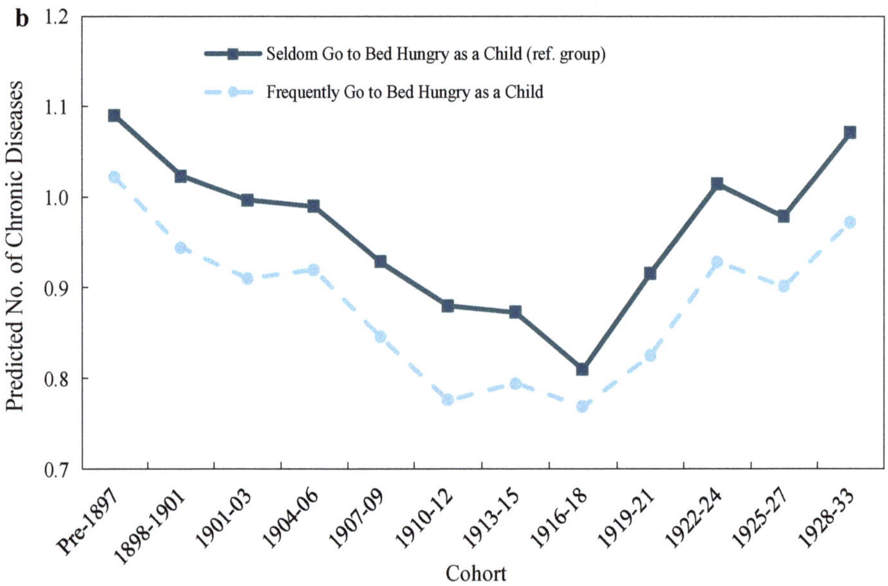

Fig. 6.9b Predicted number of chronic diseases by frequently going to bed hungry in childhood from CCREMs over Birth Cohort. (Sources: 1998–2018 Chinese Longitudinal Healthy Longevity Survey (CLHLS))

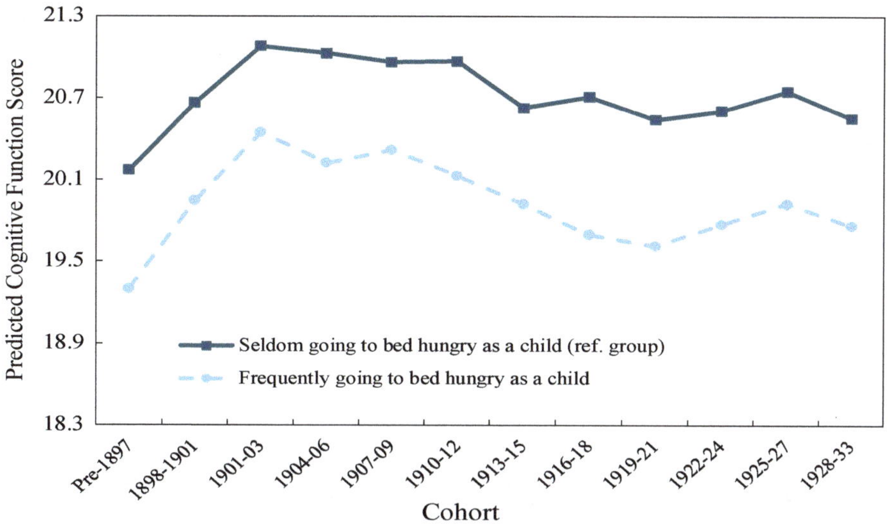

Fig. 6.10 Predicted cognitive function score by frequently going to bed hungry in childhood from CCREMs over birth cohort. (Sources: 1998–2018 Chinese Longitudinal Healthy Longevity Survey (CLHLS))

across birth cohorts. The significant cohort effects are demonstrated by a "V" shape with the 1916–18 birth cohort showing a fewer predicted number of chronic diseases than other birth cohorts.

Figure 6.10 depicts significant cohort effects drawn from regression results when predicting cognitive function scores of the oldest old. The period effects are significant. Apparently, respondents who frequently went to bed hungry in childhood show worse cognitive function than those who seldom did so. The health disparities in cognitive function remain across birth cohorts. The general pattern for both subgroups is that there is an increasing pattern for predicted cognitive function scores for birth cohorts born before 1901. Afterwards, a declining trend is found for later birth cohorts.

Regarding measures of the oldest old's subjective welling, only cohort effects are shown to be significant when predicting the respondents' positive feeling scores, which is depicted in Fig. 6.11. Neither period or cohort effects are significant when predicting the oldest old's negative feeling scores. Obviously, the oldest old who seldom went to bed hungry in childhood have higher predicted positive feeling scores as compared to those who frequently did so. Cohort variations are more apparent among the subgroup that frequently went to bed hungry as a child. These findings imply that respondents with lower SES in childhood are more likely to be influenced by cohort changes as compared to those with a better SES status.

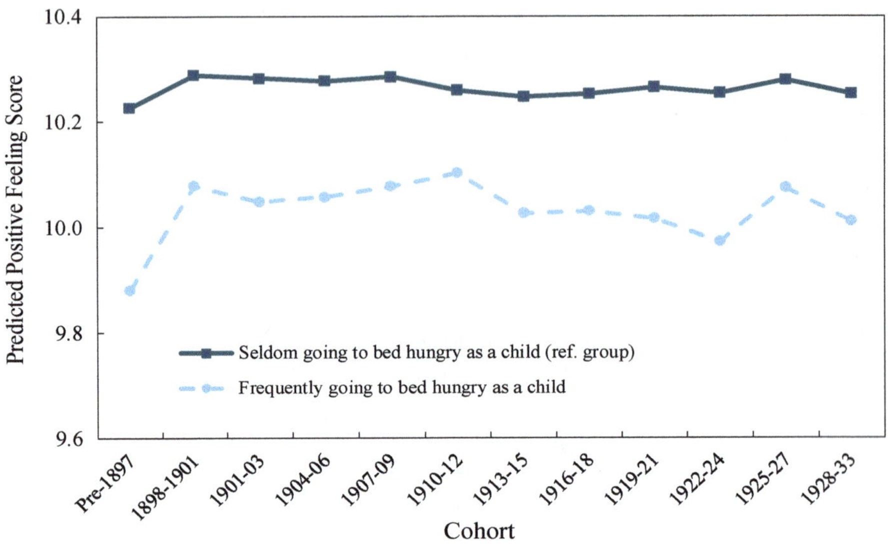

Fig. 6.11 Predicted positive feeling Score by frequently going to bed hungry in childhood from CCREMs over birth cohort. (Sources: 1998–2018 Chinese Longitudinal Healthy Longevity Survey (CLHLS))

6.4 Conclusion and Discussion

This chapter has studied the period and cohort variations when using socioeconomic factors to predict Chinese oldest old's health outcomes. This is one of the pioneer studies that have examined the compound effects of age, period and cohort at the same time. Some important findings can be drawn from this research, which has not been revealed in previous literature.

To begin with, the results fall in line with findings from cross-sectional analyses that socioeconomic factors are major determinants of the oldest old's health outcomes, controlling for demographic characteristics. The oldest old who have higher educational attainments and seldom went to bed hungry show better health outcomes except for taking the chronic disease measure into consideration. These findings suggest that higher SES is positively associated with the oldest old's health status. Individual SES status in childhood has a cumulative effect on his or her health in older ages, which supports the cumulative advantage/disadvantage theory.

The most important findings are the period and cohort variations in the linkage between SES and the oldest old's health. In general, the findings suggest that health of individuals with lower SES is more sensitive to cohort and period changes as compared to those with higher SES. The most significant period effect is shown when predicting positive feeling scores of the oldest old. There is a dip when using each SES variable to estimate the respondent's positive feeling scores. Again, this is likely to be caused by the economic hardship and *Wenchuan* earthquake that occurred in that particular year. Findings from this chapter once again highlight that social events have stronger influence on the oldest old's subjective wellbeing than physical health.

To conclude, this chapter has attempted to shed light on the period and cohort effects in the linkage between SES factors and Chinese oldest old's health. It needs to be pointed out that although CLHLS is a rich dataset that allows detailed examination of the demographic and socioeconomic determinants of health among the oldest old, there are still a number of limitations which cause measurement difficulties and lack of important variables. For example, the commonly used SES measure, income, is not studied because such information is not available in several earlier waves of the survey. In addition, the research demonstrates period and cohort effects, but the chapter has not explored why these effects exist when estimating the oldest old's health. Future research may develop more appropriate measures and investigate why such cohort and period effects have occurred to improve this current research.

References

Brasher, M. S., George, L. K., Shi, X., Yin, Z., & Zeng, Y. (2017). Incorporating biomarkers into the study of socio-economic status and health among older adults in China. *SSM Population Health, 3*, 577–585. https://doi.org/10.1016/j.ssmph.2017.07.003

Brinda, E. M., Rajkumar, A. P., Attermann, J., Gerdtham, U. G., Enemark, U., & Jacob, K. S. (2016). Health, social, and economic variables associated with depression among older people in low and middle income countries: World Health Organization study on global AGEing and adult health. *The American Journal of Geriatric Psychiatry, 24*(12), 1196–1208. https://doi.org/10.1016/j.jagp.2016.07.016

Brunello, G., Fort, M., Schneeweis, N., & Winter-Ebmer, R. (2016). The causal effect of education on health: What is the role of health behaviors? *Health Economics, 25*(3), 314–336. https://doi.org/10.1002/hec.3141

Chen, F., Yang, Y., & Liu, G. (2010). Social change and socioeconomic disparities in health over the life course in China: A cohort analysis. *American Sociological Review, 75*(1), 126–150.

Chen, C. X., Hao, X. J., Ma, Y. K., Zhang, X., Zhao, Y. N., & Li, S. X. (2015). Effects of income on physical and mental health and health self-management of the elderly. *Chinese Journal of Gerontology (In Chinese), 18*, 5292–5294.

Folstein, M. F., Folstein, S. E., & McHugh, P. R. (1975). "Mini-mentalstate": A practical method for grading the cognitive state of patients for the clinician. *Journal of Psychiatric Research, 12*, 189–198.

Harber-Aschan, L., Calderón-Larrañaga, A., Darin-Mattson, A., Hu, X., Fratiglioni, L., & Dekhtyar, S. (2020). Beyond the social gradient: The role of lifelong socioeconomic status in older adults' health trajectories. *Aging, 12*(24), 24693–24708.

Hayward, M. D., & Gorman, B. K. (2004). The long arm of childhood: The influence of early-life social conditions on Men's mortality. *Demography, 41*(1), 87–107.

Hurt, L. S., Ronsmans, C., & Saha, S. (2004). Effects of education and other socioeconomic factors on middle age mortality in rural Bangladesh. *Journal of Epidemiology and Community Health, 58*(4), 315–320. https://doi.org/10.1136/jech.2003.007351

Klopack, E. T., Thyagarajan, B., Faul, J. D., Meier, H. C. S., Ramasubramanian, R., Kim, J. K., & Crimmins, E. M. (2022). Socioeconomic status and immune aging in older US adults in the health and retirement study. *Biodemography and Social Biology, 67*(3–4), 187–202. https://doi.org/10.1080/19485565.2022.2149465

Li, D., & Bai, G. (2022). Mental health inequality in the elderly under income differences and its influencing mechanism. *Beijing Journal of Social Sciences*, (7), 108–117. https://doi.org/10.13262/j.bjsshkxy.bjshkx.220711

Liang, J., McCarthy, J., Jain, A., Krause, N., Bennett, J., & Gu, S. (2000). Socioeconomic gradient in old age mortality in Wuhan, China. *Journals of Gerontology Series B-Psychological Sciences and Social Sciences, 55*(4), S222–S233. https://doi.org/10.1093/geronb/55.4.S222

Lima-Costa, M. F., De Oliveira, C., Macinko, J., & Marmot, M. (2012). Socioeconomic inequalities in health in older adults in Brazil and England. *American Journal of Public Health, 102*(8), 1535–1541. https://doi.org/10.2105/AJPH.2012.300765

Liu, R. P., & Li, J. X. (2021). Multiple effects of childhood adversity on health of middle-aged and elderly people in China: Single, cumulative and combined effects. *Journal of Yunnan Minzu University (Philosophy and Social Sciences Edition) (in Chinese), 38*(3). https://doi.org/10.13727/j.cnki.53-1191/c.2021.03.008

Liu, R. P., & Li, J. X. (2022). Impact of childhood adversity on the health of middle-aged and elderly people. *Journal of Yunnan Minzu University (Philosophy and Social Sciences Edition) (in Chinese), 39*(6). https://doi.org/10.13727/j.cnki.53-1191/c.20221104.005

Lv, X., Li, W., Ma, Y., Chen, H., Zeng, Y., Yu, X., Hofman, A., & Wang, H. (2019). Cognitive decline and mortality among community-dwelling Chinese older people. *BMC Medicine, 17*, 63.

Marks, I. H., Thomas, H., Bakhet, M., & Fitzgerald, E. (2019). Medical equipment donation in low-resource settings: A review of the literature and guidelines for surgery and anaesthesia in low-income and middle-income countries. *BMJ Global Health, 4*(5), e001785. https://doi.org/10.1136/bmjgh-2019-001785

Meyer, O. L., Castro-Schilo, L., & Aguilar-Gaxiola, S. (2014). Determinants of mental health and self-rated health: A model of socioeconomic status, neighborhood safety, and physical activity. *American Journal of Public Health, 104*, 1734–1741.

Oftedal, S., Kolt, G. S., Holliday, E. G., Stamatakis, E., Vandelanotte, C., Brown, W. J., & Duncan, M. J. (2019). Associations of health-behavior patterns, mental health and self-rated health. *Preventive Medicine, 118*, 295–303.

Pampel, F. C., Krueger, P. M., & Denney, J. T. (2010). Socioeconomic disparities in health behaviors. *Annual Review of Sociology, 36*, 349–370.

Perry, B. C. (2002). Childhood experience and the expression of genetic potential: What childhood neglect tells us about nature and nuture. *Brain and Mind, 3*, 79–100.

Shankar, A., McMunn, A., & Steptoe, A. (2010). Health-related behaviors in older adults relationships with socioeconomic status. *American Journal of Preventive Medicine, 38*(1), 39–46. https://doi.org/10.1016/j.amepre.2009.08.026

Smith, K. V., & Goldman, N. (2007). Socioeconomic differences in health among older adults in Mexico. *Social Science & Medicine, 65*(7), 1372–1385. https://doi.org/10.1016/j.socscimed.2007.05.023

Smith, K., & Victor, C. (2019). Typologies of loneliness, living alone and social isolation, and their associations with physical and mental health. *Aging & Society, 39*, 1709–1730.

Tur-Sinai, A., & Becker, N. (2022). Risk factors and health status among older adults in Europe: A socioeconomic analysis. *Journal of Biosocial Science, 54*(5), 829–846. https://doi.org/10.1017/S0021932021000432

Wroblewska, W. (2002). Women's health status in Poland in the transition to a market economy. *Social Science & Medcine, 54*(5), 707–726. https://doi.org/10.1016/S0277-9536(01)00104-6

Xiang, X., & Wang, X. (2021). Childhood adversity and major depression in later life: A competing-risks regression analysis. *International Journal of Geriatric Psychiatry, 36*(1), 215–223. https://doi.org/10.1002/gps.5417

Yang, Y., & Land, K. C. (2006). A mixed models approach to the age-period-cohort analysis of repeated cross-section surveys, with an application to data on trends in verbal test scores. *Sociological Methodology, 36*(1), 75–97.

Yuan, H. (2021). Internet use and mental health problems among older people in Shanghai, China: The moderating roles of chronic diseases and household income. *Aging & Mental Health, 25*(4), 657–663.

Zeng, Y., Vaupel, J., Xiao, Z., Zhang, C., & Liu, Y. (2002). Sociodemographic and health profiles of the oldest old in China. *Population and Development Review, 28*, 251–273.

Zhang, L. (2021). A latent class analysis of health lifestyles and health outcomes among Chinese older adults. *Ageing & Society, 41*, 313–338.

Zimmer, Z., Chayovan, N., Lin, H.-S., & Natividad, J. (2004). How indicators of socioeconomic status relate to physical functioning of older adults in three Asian societies. *Research on Aging, 26*(2), 224–258.

Physical Activities and the Oldest Old's Health

7

In early 2000s, WHO emphasized the concept of active aging and highlighted the important influence of earlier life experience on individual health at older ages. The life course perspective began to draw more research attention in recent years. According to this perspective, the adoption of healthy lifestyles is crucial to improve older adults' health. Participation in regular and moderate physical activities can prevent functional declines in both physical and mental dimensions (World Health Organization, 2002). Under this proceeding, this chapter focuses on studying how physical activities link to Chinese oldest old's health and investigating the period and cohort effects in such a linkage.

7.1 Empirical Research on Physical Activities and Older Adults' Health

7.1.1 Cross-sectional Studies

Prior cross-sectional studies have shown that the level of physical activities has a positive effect on multiple dimensions of older adults' health. The first dimension is self-rated health (SRH). Researchers showed that physical activities can promote elders' SRH (Christodoulou et al., 2023; kailin et al., 2023; Lampek et al., 2021; Zhou et al., 2023). For instance, a study from Hungary indicated that the positive relationship between the level of sport activities and SRH was significant (Lampek et al., 2021). The second health dimension is chronic diseases. Studies showed that frequent physical activities decreased the risks of older adults having chronic diseases. For example, a study from Spain suggested that a higher physical activity level led to lower risks of hypertension among elders (Denche-Zamorano et al., 2022).

L. Zhang, *Living Longer and Healthier at Older Ages*, Advances in Studies of Aging and Health 3, https://doi.org/10.1007/978-3-031-69773-9_7

Cognitive function has also been found to be related to physical activities. Bae (2020) showed that regular exercise was positively associated with better cognitive function. Wang et al.' (2023) study also supported the hypothesis that a higher level of physical activities was linked to higher cognitive function scores. They found that the physical activity level was not statistically associated with naming and orientation, but it was a protective factor for mild cognitive impairment and visual space, attention, language, abstract ability and delayed recall scores. A Brazilian study further showed that inadequate physical activities led to lower cognitive functioning outcomes (Ihle et al., 2022).

In addition, physical activities were found to be beneficial to older adults' mental health. Zhang and Tian (2022) suggested that middle-aged and older adults with high physical activity levels had a lower risk of having depressive symptoms. Such a link was also found in a study from Greek. The study revealed that depression was associated with insufficient levels of physical activities (Christodoulou et al., 2023).

Besides levels of physical activities, prior research also documented a strong link between diverse types of physical activities and older adults' health (Koolhaas et al., 2017; W. Li et al., 2022; Linhares et al., 2022; Ryu et al., 2020). Physical activities have been generally classified into three types: aerobic, strengthening, and elastic/balancing. Researchers indicated that diverse training programs, including strength, aerobic, balance, flexibility and functional fitness, were effective in improving strength, flexibility, quality of life, bone mineral density, balance, and functional fitness and reducing the risks of falls among older women with osteoporosis (Linhares et al., 2022). Li et al. (2022) studied traditional Chinese exercise, such as "breath regulation", "body regulation" and "heart regulation" and found that all types of traditional Chinese health exercise had direct impacts on lung functioning. Such exercises also had positive effects on individual body mass index, waist circumference, blood glucose, blood lipid, and blood pressure levels.

7.1.2 Longitudinal Studies

Some studies examined physical activities and older adults' health from a longitudinal point of view. Those studies touched on the following health dimensions. The first dimension is SRH. By evaluating a cohort of community-living women aged 69–79 years of age, Papp et al. (2022) examined the correlation between self-reported changes in physical activity levels and the deterioration of the respondent's SRH. The results showed that during the studied 8.5 years, elder women had better SRH once they reached the recommended physical activity levels per week. Thus, a greater physical activity level was claimed to be an important predictor maintaining physical function and SRH among elder women.

The second health dimension relates to physical function among older adults. By conducting a longitudinal analysis, Hegde et al. (2016) found that participants who kept active over 18 years had a lower left ventricular mass index and a lower prevalence of left ventricular hypertrophy as compared to those with inadequate physical activities. In other words, higher levels of physical activities were beneficial for older adults in age-related changes in cardiac structure and function.

Researchers also conducted longitudinal studies on types of physical activities and their links to older adults' health. Cassilhas et al. (2021) studied the effects of indoor and outdoor aerobic exercise on older adults' health. They randomly assigned the participants into 4 groups: control indoor, control outdoor, indoor exercise and outdoor exercise. After 12 weeks of observation, they found that although improvements of cognitive function and brain-derived neural factor levels were observed in both indoor and outdoor exercise groups, indoor aerobic exerciser presented a better change when compared to their outdoor counterparts. Outdoor exercisers faced problems of air pollution and colder temperature, which may bring negative influenced on their health. Kim et al. (2022) analyzed the effects of exercise on indicators of bone formation after 6 months of complexed exercise, including aerobic, resistance, and balance exercises. They discovered that osteocalcin and bone alkaline phosphatase levels decreased among the control group without exercise. In contrast, the values of those indicators increased among the exercise group. The longitudinal work suggested that over time, exercise clearly played a positive role in preventing osteoporosis. A Brazilian longitudinal clinical study, spanning 12 weeks, divided 40 elderly participants into groups with depression and without depression. Each group was arranged a 12-week aquatic exercising program. The results showed that exercise had a greater positive effect on the group with depressive symptoms (Silva et al., 2019). In addition, Zhang and associates analyzed Chinese seniors from the 2001 to 2014 surveys. They found that over the studied period, the risks of frailty deceased by 26.0% among persistent aerobic exercisers. For those who never exercised, their risks of frailty increased by 59.0% (Zhang et al., 2021). These studies showed that various types of physical activities had positive influence on older adults' health over time.

In sum, prior analyses on physical activities and older adults' health have largely advanced our knowledge on health lifestyles and older adults' health. Those studies, nonetheless, are primarily cross-sectional. We know very little about the changing patterns of the oldest old's physical activities and how period and cohort play a role in shaping the oldest old's health outcomes. This chapter intends to fill the voids of prior literature by using the APC approach to analyze data from CLHLS that extends over 20 years. The chapter investigates the compound period and cohort effects when the effects of physical activities are considered.

7.2 Data, Measures and Methods

7.2.1 Data

This chapter analyses data from eight (1998, 2000, 2002, 2005, 2008, 2011–12, 2014 and 2018) waves of the 2018 Chinese Longitudinal Healthy Longevity Survey (CLHLS) which was conducted in randomly selected half of the counties/cities in 22 provinces and municipalities directly under the Central Government of China.

This chapter focuses on studying older adults aged 85 to 105 since prior literature considered persons who reported age 106 or higher as invalid cases (Zeng et al., 2002).

7.2.2 Measures

Physical activity measures used in this study are drawn from CLHLS questions and are largely consistent with those used in previous analyses. The survey asked whether the respondent exercised regularly in the past and at present. Those who exercised regularly both at present and during the past were coded as '1' and '0' if otherwise. Those who were coded '1' are considered to be physically active. This study generates a dichotomous physical activity variable which is coded as '1' if the respondent reported that he or she engaged in physical activities both in the past and at present. The variable is coded as '0' if otherwise.

Health outcomes are measured in the same way as they are measured in Chap. 4. In this part of the analysis, demographic and socioeconomic (SES) variables are also included but they are treated as control variables. Detailed description of these variables can be found in Chap. 4.

7.2.3 Methods

This analysis treats multiple waves of CLHLS data as repeated cross-section surveys. A cross-classified random-effects model (CCREM) is used to take into account the embeddedness of respondents in CLHLS surveys within a time period and a birth cohort (Yang & Land, 2006). Specifically, the model estimated fixed effects of individual-level covariates and random effects of period (or time) and birth cohort. Individuals are the level-1 unit of analysis. Periods (or time) and birth cohorts become the level-2 units of analysis. In the analysis of the CLHLS data, an individual respondent is nested in and cross-classified by a higher level context, which is period (or time) and cohort.

Beyond providing random period effects, CCREMs allow for random effects of independent variables, which is used to test for cross-period variations in the effects of key independent variable, i.e. physical activity variable. Period-specific random-effects coefficients and cohort-specific random-effects coefficients suggest potential across-period and across-cohort changes in the effects of frequently engaging in physical activities on oldest old's health status, respectively. This study specifies models of the oldest old health status as a quadratic function of age.

Several models are presented in the analysis. The first model only includes age, age-squared and the control variables as fixed effects. This model measures the overall effects of period on the oldest old's health status, controlling for age and other demographic and socioeconomic variables. The following models include the physical activity measure, controlling for the demographic and socioeconomic factors. The full individual level or level-1 equation can be expressed as:

$$Y_{ijk} = \beta_{0jk} + \beta_{1jk}AGE + \beta_{2jk}AGE^2 + \beta_{3jk}X + \sum_{p=10}^{p} \beta_p Xp + e_{ijk} \tag{7.1}$$

where Y_{ijk} represents health status of the *i*th respondent for i = 1, …, n*j* individuals within the *j*th period for j = 1, …. β_{0j} is the intercept or for respondents in period *j*; "AGE" and "AGE^2" symbolize age and age-squared, respectively. "X" indicates independent variables; i.e., the physical activity variable. β_p represents other individual-level fixed effects where *k* is the maximum number of covariates. e_{ij} represents an individual level random error term.

The level-2 model can be expressed as:

$$\beta_{0j} = \gamma_0 + p_{0j} + c_{0k} \tag{7.2}$$

This model specifies the overall mean vary from period to period and from cohort to cohort. γ_0 is the model intercept, which is the expected mean at the zero value of all level-1 variables averaged over all periods and cohorts. Here p_{0j} and c_{0k} are the residual random effects of period and cohort, respectively. The cell mean, $\beta_{0j,}$ is equal to the sum of the overall mean or intercept (γ_0), and the residual random effect of period j (p_{0j}) and the residual random effect of cohort (c_{0k}). The residual random effects of period j (p_{0j}) and cohort k (c_{0k}) examine the effects of each period and cohort on a specific health measure. When adding (7.2) to (7.1), it yields the level-1 and level-2 combined model:

$$Y_{ijk} = \gamma_0 + \beta_{0jk} + \beta_{1jk}AGE + \beta_{2jk}AGE^2 + \beta_{3jk}X + p_{0j} + c_{0k} + e_{ijk} \tag{7.3}$$

where, within each survey year j, respondent i's health is modeled as a function of his or her age, age-squared, the physical activity variable and other control variables. In the above model, γ_0 represents the fixed-effects coefficients for the physical activity variable; the symbols of p_{0j} and c_{0k} represent the period-specific and cohort-specific effects of the physical activity variable, respectively. The model tests for random effects of physical activity patterns on the respondent's health across periods and cohorts. All continuous variables are sorted and centered around their overall mean.

Since there are eight waves of survey data, survey years for the eight waves are considered as the eight periods. I have developed 13 birth cohorts in the analyses (pre-1897, 1898–1900, 1901–03, 1904–06, 1907–09, 1910–12, 1913–15, 1916–18, 1919–21, 1922–24, 1925–27, 1928–30 and 1931–33). The 1934 and later birth cohorts are dropped from the analyses since those cohorts do not fall into the age range of 85 to 105. When showing descriptive results, all 13 birth cohorts are shown in graphs. When conducting regression analyses, the 1928–30 and 1931–33 birth cohorts are combined into one birth cohort considering the number of respondents who were born in and after 1928 is limited.

7.3 Results

7.3.1 Descriptive Statistics

The chapter first describes the changing patterns of the oldest old's physical activities when time period and birth cohorts are considered. The study relies on graphs to show how physical activity patterns among Chinese oldest old have changed by time period and across birth cohorts (see Fig. 7.1). According to Fig. 7.1, percentages of the oldest old who engaged in physical activities both in the past and at present for various birth cohorts show a general declining trend with the values changing from 28.2% in 1998 to 4.8% in 2014. It makes sense because with aging, older adults became less active as compared to previous years. There is a slight increase in such percentages in year 2014 for several birth cohorts.

Cohort wise, it is shown that percentages of the respondents who participated in physical activities both previously and at present is increasing among younger birth cohorts. The finding indicates that younger birth cohorts are more active as compared to previous birth cohorts.

7.3.2 Regression Results Based on APC Analyses

This section moves to the presentation of regression results from CCREMs. Previous chapters have already shown that the relationships between age and health measures are linear. That is, adjusting for time period and birth cohort variations,

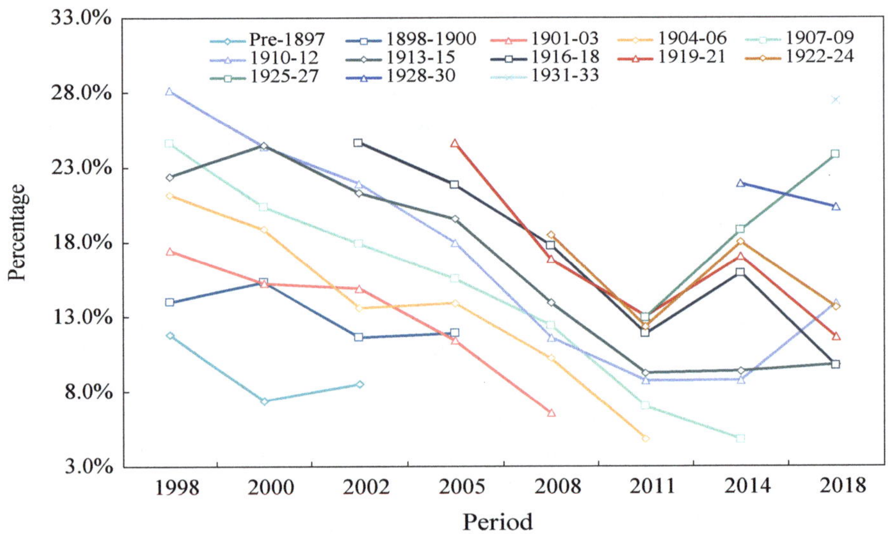

Fig. 7.1 Percentage of respondents who participated in physical activities both in the past and at present (Unweighted). (Sources: 1998–2018 Chinese Longitudinal Healthy Longevity Survey (CLHLS))

with age increasing, the level of SRH increases but the levels of other health dimensions decline. This chapter focuses on estimating the random effects in terms of residual variance components at level-2 and test if significant period and cohort effects exist when using physical activity variables to predict the health outcome measures, controlling for demographic and SES variables.

Table 7.1 presents regression results from the CCREMs when including control variables, i.e., demographic and socioeconomic factors, as well as the physical activity variable in the analyses. The results show that men on average show better SRH and a lower number of chronic diseases, they also have better subjective wellbeing as compared to women. Urban oldest old also show better subjective wellbeing but a greater number of chronic diseases than their rural counterparts. Married individuals show better cognitive function and subjective wellbeing, but they have worse SRH and more chronic diseases. Higher SES, measured by schooling years and going to bed hungry in childhood, leads to better SRH, cognitive function and subjective wellbeing. But higher SES is also found to be associated with a greater number of chronic diseases.

Since the next chapter examines the association between leisure activities and the oldest old's health, this chapter shows the CCREM results when including both physical activity and leisure activity variables in the first level. For the fixed effects in level 1, results show that physical activity variable has significant effects on the oldest old's health. As Table 7.1 shows, as compared to oldest old who did not participate in physical activities, those who remained physically active both in past and at present show better SRH, cognitive function and subjective wellbeing.

The next part of the analyses focuses on the random effects which show significant period and cohort changes in physical activity patterns net of age effects and other factors. The significant period and cohort changes are presented in graphs. Figure 7.2a shows the oldest old who exercised both in the past and at present have better predicted SRH as compared to those who did not do so. The reference group includes those who never exercised, who exercised either in the past or at present. The predicted SRH generally shows a declining pattern over time with a slight increase in years 2008 and 2018. As to the cohort effects shown in Fig. 7.2b, SRH of those who engaged in physical activities constantly is more likely to be affected by cohort changes.

Figures 7.3a and 7.3b depict significant period and cohort effects, respectively, when linking physical activity to the predicted number of chronic diseases, controlling for demographic and socioeconomic covariates as well as the period and cohort effects. As it is shown in Fig. 7.3a, prior to year 2008, those who did not participate in physical activities constantly tend to report a greater number of chronic diseases. After 2008, the reverse pattern seems to be true. That is, the oldest old who exercised both in the past and at present report more chronic diseases. Over time, the predicted number of chronic diseases is increasing for both subgroups.

The cohort effects are shown in Fig. 7.3b. Clearly, the disparities in predicted numbers of chronic diseases are minimal with those who participated in physical activities both in the past and at present showing a slightly fewer number of chronic diseases. The significant cohort effects are shown in a V-shape trend in the predicted

Table 7.1 Regression results of health outcome measures on physical activity and other control variables from CCREM models: Chinese aged 85–105

	Self-rated health		Chronic disease		Cognitive function		Positive feeling		Negative feeling	
Variables	Coef.	P Value	Coef.	P Value	Coef.	P Value	Coef.	P Value	Coef.	P Value
Intercept	3.212	<0.001	0.879	<0.001	18.376	<0.001	9.590	<0.001	7.911	<0.001
Control variables										
Age	0.007	<0.001	−0.014	<0.001	−0.316	<0.001	−0.002	0.447	−0.016	<0.001
Age^2	0.000	0.120	0.000	0.178	0.000	0.691	−0.001	0.102	0.000	0.319
Sex (ref. = female)	0.033	<0.001	−0.076	<0.001	1.412	<0.001	0.013	0.617	−0.158	<0.001
Residence (ref. = rural)	0.000	0.989	0.281	<0.001	−0.018	0.780	0.217	<0.001	−0.184	<0.001
Being married	−0.066	<0.001	0.074	<0.001	0.403	<0.001	−0.008	0.806	−0.379	<0.001
Schooling years	−0.002	0.158	0.047	<0.001	0.159	<0.001	0.033	<0.001	−0.021	<0.001
Often going to bed hungry as a child (ref. = yes)	0.043	<0.001	0.018	0.140	0.534	<0.001	0.161	<0.001	−0.194	<0.001
Activity variables										
Physical activity	0.296	<0.001	−0.016	0.743	1.671	0.002	0.754	<0.001	−0.560	<0.001
Leisure activity	0.279	<0.001	−0.038	0.002	3.045	<0.001	0.828	<0.001	−0.662	<0.001
Random effect component										
Period										
Intercept	0.011	0.034	0.058	0.033	0.314	0.037	0.041	0.034	0.105	0.032
Physical activity	0.003	0.080	0.016	0.061	0.220	0.103	0.033	0.059	0.057	0.046
Cohort										
Intercept	0.000	0.171	0.007	0.019	0.109	0.027	0.002	0.159	0.002	0.151
Physical activity	0.001	0.144	0.000	0.481	0.986	0.033	0	–	0.001	0.377
AIC	122,284.5		167,912.3		302,850.0		213,217.4		201,524.3	
BIC	122,274.5		167,902.3		302,840.0		213,209.4		201,514.3	

Source: 1998–2018 Chinese Longitudinal Healthy Longevity Survey (CLHLS)

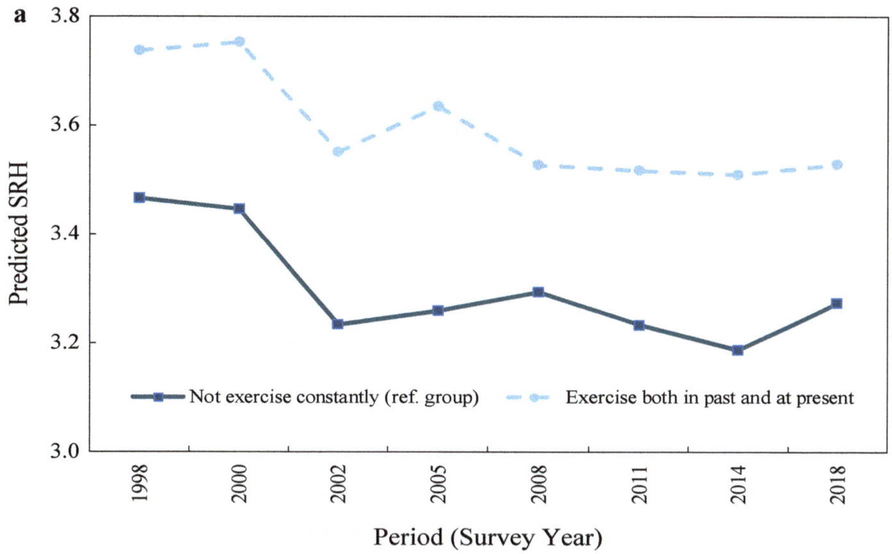

Fig. 7.2a Predicted SRH by physical activity over time period. (Sources: 1998–2018 Chinese Longitudinal Healthy Longevity Survey (CLHLS))

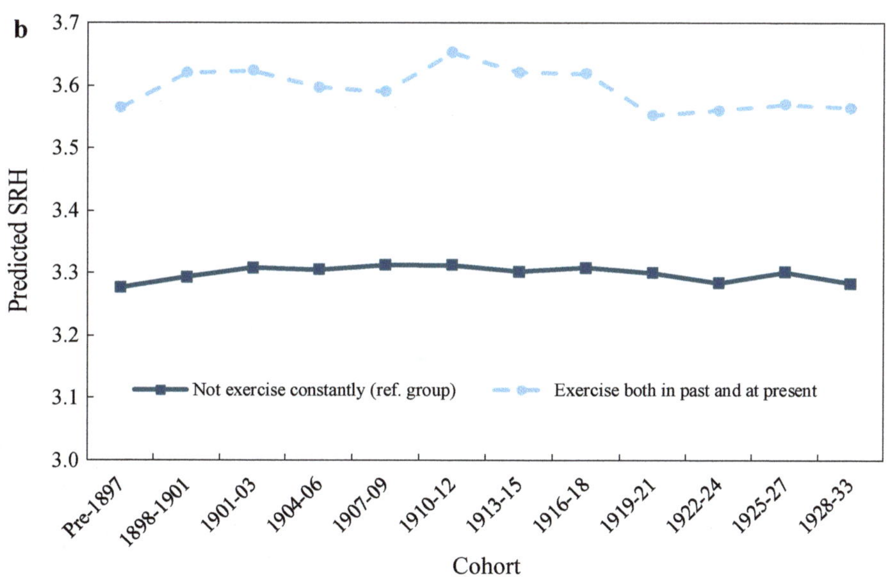

Fig. 7.2b Predicted SRH by physical activity over birth cohort. (Sources: 1998–2018 Chinese Longitudinal Healthy Longevity Survey (CLHLS))

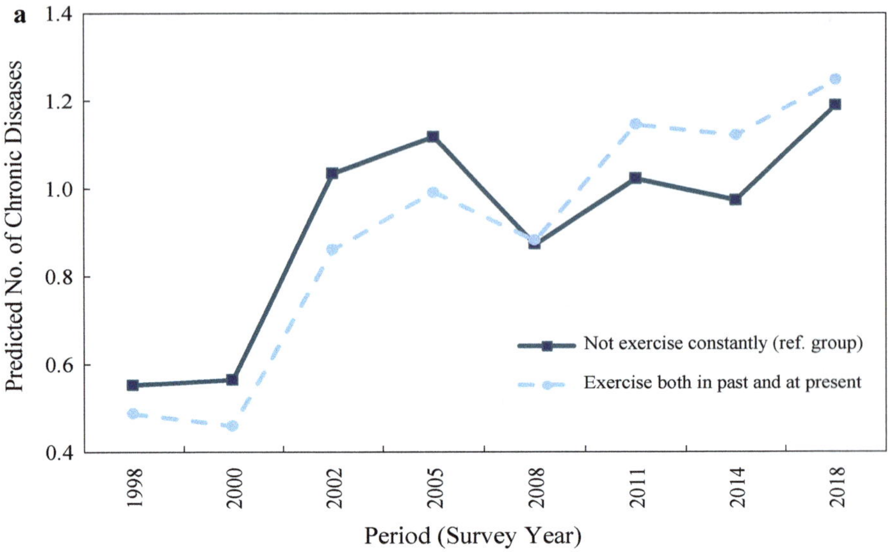

Fig. 7.3a Predicted number of chronic diseases by physical activity over time period. (Sources: 1998–2018 Chinese Longitudinal Healthy Longevity Survey (CLHLS))

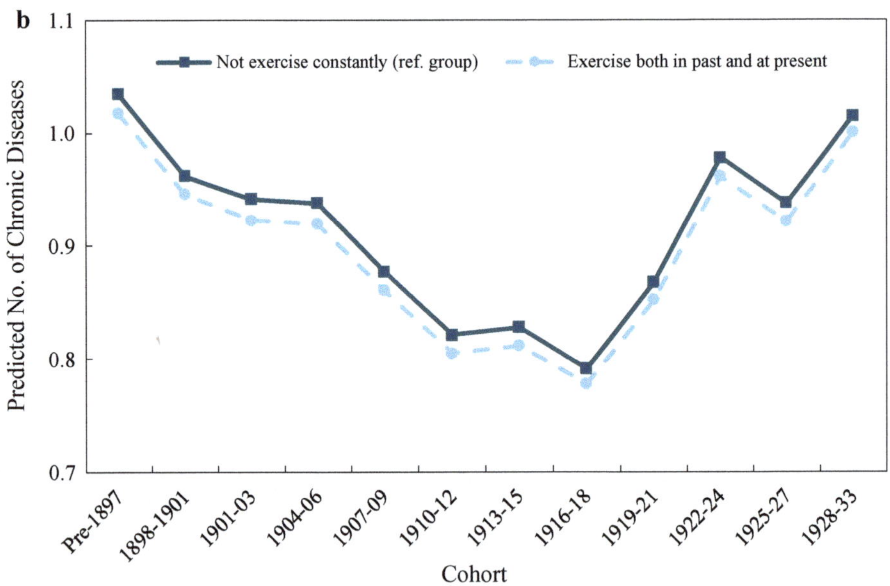

Fig. 7.3b Predicted number of chronic diseases by physical activity over birth cohort. (Sources: 1998–2018 Chinese Longitudinal Healthy Longevity Survey (CLHLS))

number of chronic diseases for both subgroups. The 1916–18 birth cohort has the lowest predicted number of chronic diseases as compared to other younger and older birth cohorts.

When using the physical activity variable to predict cognitive function, significant period and cohort effects are shown, which are described by Figs. 7.4a and 7.4b, respectively. Figure 7.4a shows that during 1998 to 2018, the predicted cognitive function scores for the oldest old who exercised both in the past and at present are higher than cognitive function scores for the group that did not do so (19.6 to 22.0 vs.18.6 to 20.1). Over time, the predicted cognitive function scores first show a decreasing trend and an overall increasing trend is observed after year 2000. These findings suggest that the cognitive function among Chinese oldest old is getting better over time with physically more active group showing better cognitive function. The cohort effects shown in Fig. 7.4b demonstrate that the health disparities in cognitive function between constant and non-constant exercising groups are shrinking across birth cohorts. And it is very interesting to see that the predicted cognitive function scores for the constant exercising group is declining across birth cohorts. I suspect that this is probably because that the positive effect of exercising on health is decreasing, whereas other social, cultural, behavior or even medical factors may have played a more important role in determining the oldest old's health among younger birth cohorts. Of course, further investigation is needed to justify this explanation.

Only significant period effects are observed when predicting positive feeling scores of the respondents, which is shown in Fig. 7.5. The figure indicates that the oldest old who engaged in physical activities both in the past and at present have

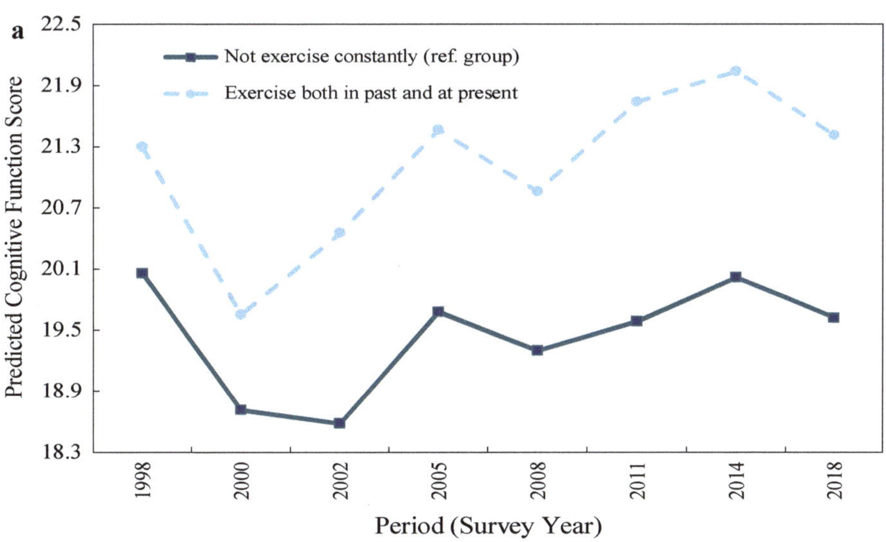

Fig. 7.4a Predicted cognitive function score by physical activity over time period. (Sources: 1998–2018 Chinese Longitudinal Healthy Longevity Survey (CLHLS))

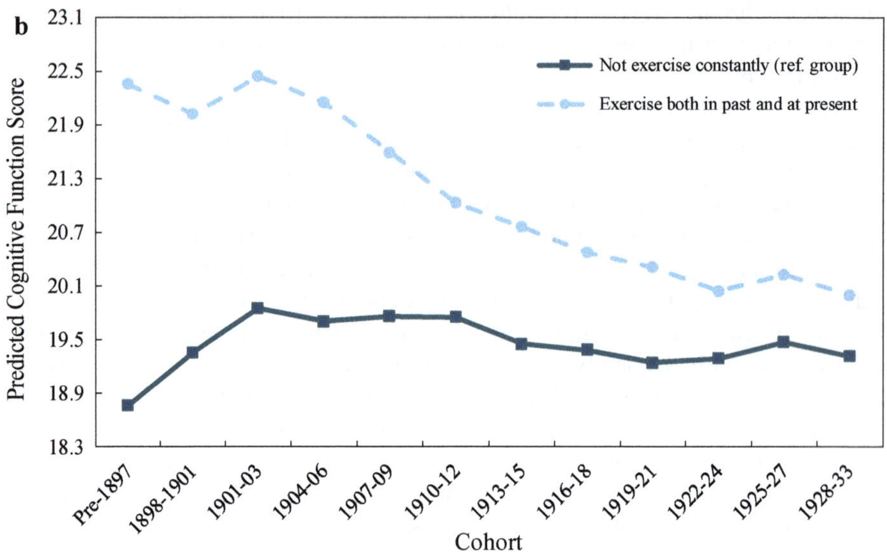

Fig. 7.4b Predicted cognitive function score by physical activity over Birth Cohort. (Sources: 1998–2018 Chinese Longitudinal Healthy Longevity Survey (CLHLS))

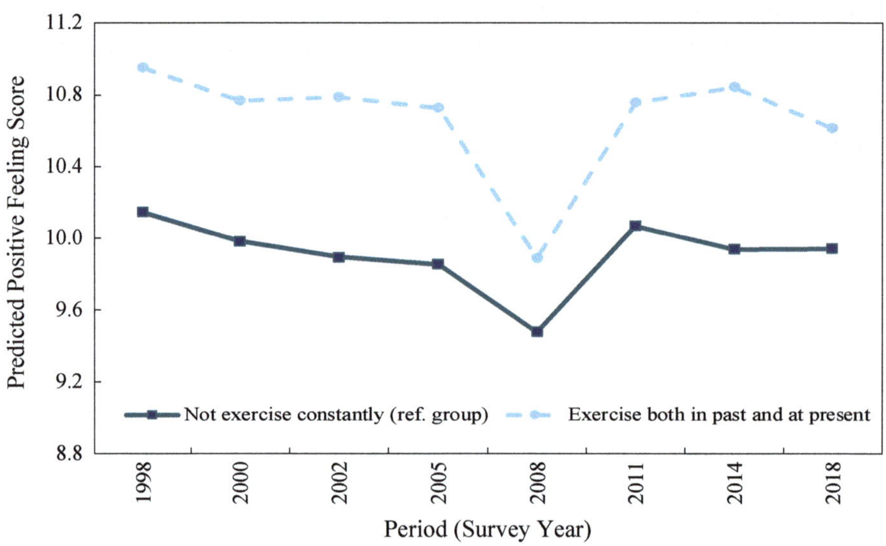

Fig. 7.5 Predicted Positive Feeling Score by Physical Activity over Time Period. (Sources: 1998–2018 Chinese Longitudinal Healthy Longevity Survey (CLHLS))

higher positive feeling scores as compared to the less active group, i.e., the reference group. There is a significant dip in predicted positive feeling scores in year 2008, indicating a strong negative effect of social events on the oldest old's subjective wellbeing.

The CCREM results show significant period and cohort effects when using physical activity variable to predict negative feeling scores (see Figs. 7.6a and 7.6b). It is apparent that those who exercised both in the past and at present show lower negative feeling scores as compared to the reference group. It again supports the argument that exercising benefits the oldest old's physical and mental health. Cohort effects are demonstrated by fluctuations in predicted negative feeling scores across birth cohorts. Figure 7.6b reveals that the gaps in negative feeling scores between two subgroups remain across birth cohorts.

7.4 Conclusion and Discussion

This chapter has concentrated on exploring the associations between Chinese oldest old's health and their physical activity patterns. Although previous studies have demonstrated that physical activity participation has significantly positive effects on older adults' health, most studies have examined the subject from a cross-sectional point of view. Very little research has been done to investigate the compound effects of time and cohort on the oldest old's health due to their physical activity patterns.

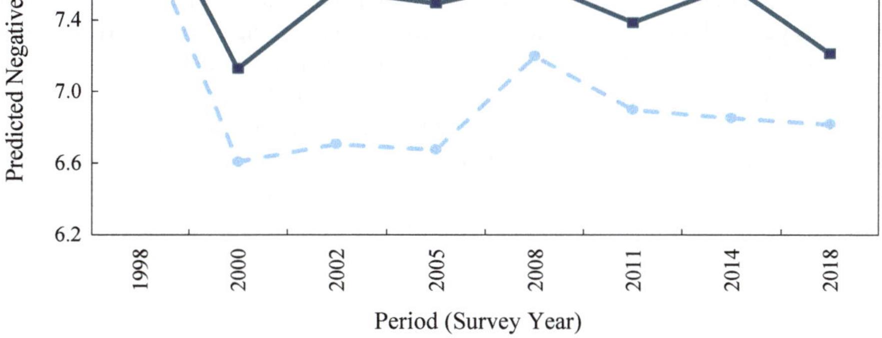

Fig. 7.6a Predicted Negative Feeling Score by Physical Activity over Time Period. (Sources: 1998–2018 Chinese Longitudinal Healthy Longevity Survey (CLHLS))

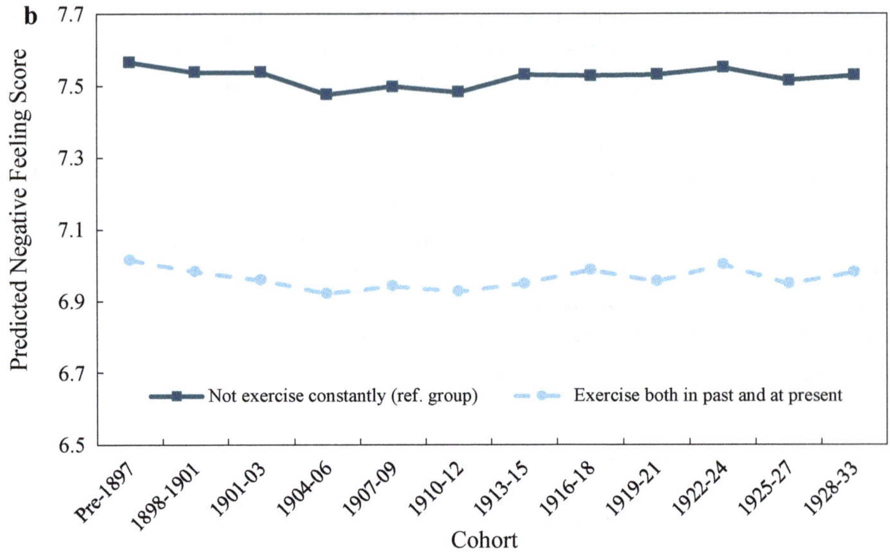

Fig. 7.6b Predicted Negative Feeling Score by Physical Activity over Birth Cohort. (Sources: 1998–2018 Chinese Longitudinal Healthy Longevity Survey (CLHLS))

This chapter contributes the literature by studying the changing patterns of Chinese oldest old's physical activity patterns over time and across birth cohorts. Through constructing CCREMs to evaluate the time-series data on Chinese oldest old's health that span 20 years, this chapter reveals new findings that have not been documented in prior literature.

To start with, the results show period and cohort variations in Chinese oldest old's physical activity patterns during the studied 20-year span. Based on the descriptive results, percentages of the oldest old who exercised both in the past and at present show a declining trend over time. This makes sense because as time goes by, the respondents were getting older and they became less active as compared to previous years. But such percentages seem to increase among younger birth cohorts, indicating that younger birth cohorts are more physically active than older ones.

In addition, the health disparities predicted by the oldest old's physical activity variable also vary by period and cohort. When physical activity covariate is used to predict Chinese oldest old's health status, the results from CCREMs show that the period and cohort effects are significant and the general trends are largely consistent with the ones stated in previous chapters with only age and age squared terms being included in CCREMs. The main period and cohort effects can be summarized as follows:

Firstly, when using the physical activity variable to predict the oldest old's health, those who exercised both in the past and at present have better SRH, cognitive function, subjective wellbeing and a fewer number of chronic diseases than the reference group, i.e., the more sedentary group. Significant period and cohort effects are shown here. Specifically, the predicted cognitive function tends to increase over time. The positive feeling scores show a significant drop in year 2008; and the predicted SRH shows a declining trend over time.

Secondly, cohort changes are observed when predicting SRH, cognitive function, chronic diseases and negative feeling scores. The chapter has not investigated why such period and cohort effects existed. It certainly deserves further enquiry why the 1916–18 birth cohort shows a fewer number of chronic diseases than the other birth cohorts when it is linked to physical activities. Another interesting finding that needs to be emphasized here is that the cognitive function advantage among the more active group is declining across birth cohorts. Does it mean that the positive effect of exercising on the oldest old's health has been decreasing across birth cohorts? I recommend readers to interpret this finding with caution. This is because the health disparities between the more active and sedentary groups persist across birth cohorts when health is measured by SRH, chronic diseases and subjective well-being. Thus, I would suspect that the positive influence of exercising on other dimensions of the oldest old's health remains, but such influence on the oldest old's cognitive function may have diminished. Other social, cultural, behavioral and medical factors may have played a stronger role in benefiting the oldest old's cognitive function among younger birth cohorts. Those factors certainly deserve further investigation in future research.

In conclusion, this chapter has attempted to investigate the period and cohort effects in the relationship between Chinese oldest old's physical activities and their health status. This research has showed that younger birth cohorts are more active in physical activity levels. The health disparities between the sedentary and more active groups revealed by this study have again emphasized the importance of active lifestyles in maintaining the oldest old's healthy status. Increasing physical activity level is certainly essential in promoting healthy aging.

References

Bae, S. M. (2020). The association between health-related factors, physical and mental diseases, social activities, and cognitive function in elderly Koreans: A population-based cross-sectional study. *Psychogeriatrics, 20*(5), 654–662. https://doi.org/10.1111/psyg.12561

Cassilhas, R. C., De Sousa, R. A. L., Caxa, L., Viana, V., Meeusen, R., Gonçalves, F. L., Diniz e Magalhães, C. O., Tufik, S., Dias Peixoto, M. F., Monteiro Junior, R. S., & de Mello, M. T. (2021). Indoor aerobic exercise reduces exposure to pollution, improves cognitive function, and enhances BDNF levels in the elderly. *Air Quality, Atmosphere & Health, 15*(1), 35–45. https://doi.org/10.1007/s11869-021-01083-x

Christodoulou, E., Pavlidou, E., Mantzorou, M., Koutelidakis, A., Vadikolias, K., Psara, E., Vorvolakos, T., Antasouras, G., Serdari, A., Vasios, G., & Giaginis, C. (2023). Depression is associated with worse health-related quality of life, lower physical activity levels, and inadequate sleep quality in a Greek elderly population. *Psychology, Health & Medicine, 28*(9), 2486–2500. https://doi.org/10.1080/13548506.2023.2221446

Denche-Zamorano, A., Perez-Gomez, J., Mendoza-Munoz, M., Carlos-Vivas, J., Oliveira, R., & Brito, J. P. (2022). Risk of hypertension and use of antihypertensive drugs in the physically active population under-70 years old-Spanish health survey. *Healthcare (Basel), 10*(7), 1283. https://doi.org/10.3390/healthcare10071283

Hegde, S. M., Goncalves, A., Claggett, B., Evenson, K. R., Cheng, S., Shah, A. M., Folsom, A. R., & Solomon, S. D. (2016). Cardiac structure and function and leisure-time physical activity in the elderly: The atherosclerosis risk in communities study. *European Heart Journal, 37*(32), 2544–2551. https://doi.org/10.1093/eurheartj/ehw121

Ihle, A., Gouveia, E. R., Gouveia, B. R., Marques, A., Marconcin, P., Nascimento, M. M., Haas, M., Jurema, J., Tinoco, M. A., & Kliegel, M. (2022). The relation of having experienced a fall in the past to lower cognitive functioning in old age is mediated via less physical activity engagement as cognitive reserve contributor. *Biology (Basel), 11*(12), 1754. https://doi.org/10.3390/biology11121754

Kailin, G., Shiqiang, W., Dan, L., Zhihan, X., Yijie, W., & Shaokun, W. (2023). Correlation between physical activity and frailty in urban elderly of China (in Chinese). *Chinese Journal of Prevention and Control of Chronic Diseases, 31*(3), 161–164. https://doi.org/10.16386/j.cjpccd.issn.1004-6194.2023.03.001

Kim, A. R., Lee, S.-E., Shim, Y.-J., & Choi, S.-W. (2022). The effect of 6-month complex exercise on serum bone metabolism: Focused on the elderly over 75 years old. *Applied Sciences, 12*(22), 11373. https://doi.org/10.3390/app122211373

Koolhaas, C. M., Dhana, K., van Rooij, F. J. A., Schoufour, J. D., Hofman, A., & Franco, O. H. (2017). Physical activity types and health-related quality of life among middle-aged and elderly adults: The Rotterdam study. *Journal of Nutrition, Health & Aging, 22*(2), 246–253. https://doi.org/10.1007/s12603-017-0902-7

Lampek, K., Csoka, L., Hegedus, R., Zrinyi, M., & Torocsik, M. (2021). Sports activities of 60 above Hungarian elderly-explaining and predicting impact of exercise on health. *BMC Public Health, 21*(Suppl 1), 1863. https://doi.org/10.1186/s12889-020-09974-x

Li, W., Song, Y., Xiang, Q., Wang, X., Wei, X., & Fan, T. (2022). Effect of different conditioning methods of traditional Chinese health exercise on lung function in healthy middle-aged and elderly people: Study protocol for a randomized controlled trial. *Trials, 23*(1), 8. https://doi.org/10.1186/s13063-021-05980-5

Linhares, D. G., Borba-Pinheiro, C. J., Castro, J. B. P., Santos, A., Santos, L. L. D., Cordeiro, L. S., Drigo, A. J., Nunes, R. A. M., & Vale, R. G. S. (2022). Effects of multicomponent exercise training on the health of older women with osteoporosis: A systematic review and meta-analysis. *International Journal of Environmental Research and Public Health, 19*(21), 14195. https://doi.org/10.3390/ijerph192114195

Papp, M. E., Grahn-Kronhed, A. C., Rauch Lundin, H., & Salminen, H. (2022). Changes in physical activity levels and relationship to balance performance, gait speed, and self-rated health in older Swedish women: A longitudinal study. *Aging Clinical and Experimental Research, 34*(4), 775–783. https://doi.org/10.1007/s40520-021-02016-5

Ryu, M., Lee, S., Gym, H., Baek, W. C., & Kimm, H. (2020). Analysis of association of occupational physical activity, leisure-time physical activity, and sedentary lifestyle with hypertension according to the adherence with aerobic activity in women using Korea National Health and nutrition examination survey 2016-2017 data. *International Journal of Hypertension, 2020*, 8943492. https://doi.org/10.1155/2020/8943492

Silva, L. A. D., Tortelli, L., Motta, J., Menguer, L., Mariano, S., Tasca, G., Silveira, G. B., Pinho, R. A., & Silveira, P. C. L. (2019). Effects of aquatic exercise on mental health, functional autonomy and oxidative stress in depressed elderly individuals: A randomized clinical trial. *Clinics (São Paulo, Brazil), 74*, e322. https://doi.org/10.6061/clinics/2019/e322

Wang, X., Zhang, J., Chen, C., Lu, Z., Zhang, D., & Li, S. (2023). The association between physical activity and cognitive function in the elderly in rural areas of northern China. *Frontiers in Aging Neuroscience, 15*, 1168892. https://doi.org/10.3389/fnagi.2023.1168892

World Health Organization. (2002). Active ageing: A policy framework. *The Aging Male, 5*(1), 1–37. https://doi.org/10.1080/tam.5.1.1.37

Yang, Y., & Land, K. C. (2006). A mixed models approach to the age-period-cohort analysis of pepeated cross-section surveys, with an application to data on trends in verbal test scores. *Sociological Methodology, 36*(1), 75–97.

Zeng, Y., Vaupel, J., Xiao, Z., Zhang, C., & Liu, Y. (2002). Sociodemographic and health profiles of the oldest old in China. *Population and Development Review, 28*, 251–273.

Zhang, Y., & Tian, Y. (2022). The relationship between physical activity and depressive symptoms in middle-aged and elderly people controlling for demographic and health status variables. *Sustainability, 14*(21), 13986. https://doi.org/10.3390/su142113986

Zhang, Y. J., Yao, Y., Zhang, P. D., Li, Z. H., Zhang, P., Li, F. R., Wang, Z. H., Liu, D., Lv, Y. B., Kang, L., Shi, X. M., & Mao, C. (2021). Association of regular aerobic exercises and neuromuscular junction variants with incidence of frailty: An analysis of the Chinese longitudinal health and longevity survey. *Journal of Cachexia, Sarcopenia and Muscle, 12*(2), 350–357. https://doi.org/10.1002/jcsm.12658

Zhou, Y., Huang, Y., Zhang, A., Yin, G., & Hu, H. (2023). Determinants of self-rated health among elderly patients with hypertension: A cross-sectional analysis based on the Chinese longitudinal healthy longevity survey. *Clinical and Experimental Hypertension, 45*(1), 2224942. https://doi.org/10.1080/10641963.2023.2224942

Leisure Activities and the Oldest Old's Health

8

Prior studies have documented a strong link between leisure activities and older adults' health. A number of theories have been proposed to explain the relationship between active aging and older adults' health status. For instance, the *"use it or lose it" theory* considered human body as a machine and effective exercises can maintain normal function. The theory suggested that older adults who engage in certain leisure activities may receive training in relevant functions, such as cognition, locomotive function, memory function, logical function, et al. (Jung et al., 2017; Bhattacharya et al., 2023). The *hierarchy of needs theory* also proposed five aspects of human needs, including psychological needs, safety, belonging, self-esteem, and self-actualization. Different types of leisure activities have been claimed to be able to meet hierarchical needs of individuals to a certain extent, which in turn benefits one's health (Du, 2022; Bae, 2020). The *planned behavior theory* further argued that human behaviors are rational and goal directed (Li et al., 2019). Everyone has the need of development, including self-determination, capability, and affiliation. Leisure activities can be considered as planned behaviors which somewhat prevent seniors' from being isolated and increase the frequency of interaction between older adults and others (Fan et al., 2023). The above theories have provided theoretical guidance for empirical analyses when linking leisure activities and individual health. This chapter aims to investigate if period and cohort effects exist when leisure activities influence the oldest old's health outcomes.

8.1 Empirical Research on Leisure Activities and Older Adults' Health

8.1.1 Cross-sectional Studies

The broad use of the term 'physical activity' is often equated with exercise. But the definition of 'leisure activity' is vague in the literature and there is terminological

L. Zhang, *Living Longer and Healthier at Older Ages*, Advances in Studies of Aging and Health 3, https://doi.org/10.1007/978-3-031-69773-9_8

confusion. Even with the confusion, there appears to be some agreements in the definition of leisure activity. Most studies defined it as activities completed during leisure time (Zhang et al., 2017). Such a broad definition, however, brings a question that whether physical activities in leisure time should be considered as physical activities or leisure activities. To clarify the boundary, leisure activity has been further defined as 'leisure-time non-exercise activities', including gardening work, raising domestic animals, reading books or newspaper, playing cards et al. (Zhu et al., 2017; Li & Zhenhui, 2017; Silva, 2021).

A number of theories have been proposed discussing leisure activity participation and its benefits. One of the major theories is the *continuity theory of aging,* which emphasizes that continuous participation in leisure activities of the same kind is a good way to cope with and adapt to aging. Based on this theory, people do not really "change" as they age; they are becoming"more" of what they have always been. This theory proposes that maintaining patterns of thought, behavior, and habits is the best and most commonly used strategy for adapting to changes during the aging process, such as experiencing retirement, changing social networks, having physical restrictions, et al. It tries to balance the coexistence of changes and continuity (Atchley, 1999).

The second theory is *activity theory of aging* which argues that successful aging occurs with older adults staying active and maintaining social interactions. Socially active may delay the aging process. Nimrod and Kleiber (2007) further put forward the *innovation theory of successful aging* which argues that innovation in activities can be enjoyable and meaningful, which may lead to successful aging. The main ideas of successful aging paradigm include life satisfaction, longevity, freedom from disability, active engagement with life, and independence (Zhang & Zhang, 2021).

Researchers have applied different categories when studying leisure activity variations. Most of the classification falls into the following two or three groups, depending on whether or not they are (1) social, intellectual, or sporty, or (2) individual or collective. One commonly used classification is: physical, mental, and social (Zhang & Zhang, 2021).

Cross-sectional studies of leisure activities and older adults' health have focused on the following subjects: first, the effects of leisure activities on older adult's SRH (Silva, 2021; Confortin et al., 2015; Araújo et al., 2011; Adjei & Brand, 2018). A study on Portuguese seniors found that poor SRH was inversely associated with leisure-time activities (Araújo et al., 2011). Adjei and Brand (2018) observed a positive association between time devoted to leisure housework activities and health status among older men and women. Compared to those who spent 1–3 h per day on housework, seniors who spent 3–6 h per day had higher odds of reporting good health.

The second subject touched on older adults' cognitive function (Li et al., 2021; Bae, 2020; Jung et al., 2017). A study of Korean sample found that the frequency of attending religious activities, engaging in social activities, and participating in leisure activities significantly improved cognitive function of seniors (Bae, 2020). Zhang et al. (2022) also suggested that the habit of reading had a significantly positive effect on cognition function among Chinese elders.

Prior literature also documented a strong link between psychological well-being and seniors' leisure activity levels (Chang et al., 2022; Michele et al., 2019; Li et al., 2019; Zhang et al., 2022; Zhao, 2022). Li et al. (2019) found that leisure activity participation, especially Tai Chi, was vital for older adults' subjective wellbeing. Zhang et al. (2022) further showed that the more often older adults read books and newspapers, the more frequently they participated in social activities. Consequently, frequent social activity participation can significantly improve older adults' SRH and alleviate their depressive symptoms.

Besides the above studies, previous studies also explored how different forms of leisure activities affected seniors' health status. Bum et al. (2020) differentiated leisure activities to three types based on their sociality levels, which are active, passive, and social activities. Their results revealed that passive leisure activity participants had relatively lower scores of health measures than the other two groups. Active leisure activity participants had the highest physical health scores. But there were no statistically significant differences in levels of psychological health and happiness among seniors who participated in active, passive, and social leisure activities. Based on *Hierarchy of Needs Theory*, prior research also classified leisure activities into two categories: voluntary, and cultural and sports. Voluntary leisure activities were found to be able to alleviate difficulties in instrumental activity of daily living (IADL); cultural and sports related leisure activities were found to be effective in reducing depression and difficulties in activity of daily living (ADL) (Du, 2022).

8.1.2 Longitudinal Studies

As compared to cross-sectional studies, longitudinal analyses on leisure activities and older adults' health are sparse. The existing studies touched on the following health dimensions under this topic: first, the changing correlations between leisure activities and difficulties in IADL and ADL among older adults. Komatsu et al. (2019) studied 3,241 seniors aged 65 and over who participated in leisure activities and who didn't during a 5-year period. They found that 10.4% of the participants developed an IADL decline after 5 years. Compared with seniors without physical and cultural leisure activities, those who engaged in both physical and cultural leisure activities showed significantly lower risks of having IADL decline. This study suggested that engagement in leisure activities was associated with a significantly lower risk of having IADL decline over time (Komatsu et al., 2019). A prospective cohort study examined how leisure activities had an impact on 80-year-old individuals' ADL function over time. During the studied period, older adults participating in more physical leisure activities were observed to have a lower risk of having bathing, toileting and transferring difficulties as compared with those who engaged in intellectual and balancing leisure activities (Xiaoyun et al., 2022).

Previous studies also examined leisure activities and older adults' cognitive function changes during a specific time frame. A recent study explored cognitive function decline among Chinese seniors who participated in intellectual leisure activities and who engaged in social activities. The findings revealed that as compared to those who participated in social activities, seniors participating in intellectual activities showed a better improvement in cognitive function over time (Li et al., 2022). Xu and associates studied 3,544 participants from CLHLS datasets who had cognitive function issues during the 2002 to 2014 period. They found that during the studied 12 years, 49.2% of participants' cognitive function recovered to normal situation. They emphasized that older adults who maintained high frequency of leisure activities had a higher reversion rate of mild cognitive impairment (Xu et al., 2022).

Moreover, exisitng studies investigated the links between various types of leisure activities and senior's health status over time. Hansdottir et al. (2022) studied cognitive function decline of older adults in a five-year studied time frame. By categorizing leisure activities as cognitive individual activity, social activity, cognitive group activity, creative activity and physical activity, they found that individuals participating in creative activities exhibited lower risks of cognitive impairment. They also showed that the frequency of meeting with friends and relatives was associated with a reduced likelihood of cognitive impairment during the studied period.

Overall, the existing analyses on leisure activities and older adults' health have expanded the literature on health lifestyles and older adults' health. But most studies have been restricted to a cross-sectional scope and it has hardly explored how leisure activities have influenced older adults' health when the period and cohort components are taken into consideration. Thus, this chapter aims to improve the existing literature by analyzing data from CLHLS that extends over 20 years by constructing CCREMs. The chapter focuses on estimating the period and cohort effects when considering the effects of leisure activities on the oldest old's health. This is an innovative study of Chinese oldest old's health status by performing a systematic age, period, and cohort analysis to investigate how Chinese oldest old's health has been affected by leisure activities over time and across birth cohorts.

8.2 Data, Measures and Methods

8.2.1 Data

This chapter analyses data from eight (1998, 2000, 2002, 2005, 2008, 2011–12, 2014 and 2018) waves of the 2018 Chinese Longitudinal Healthy Longevity Survey (CLHLS) which was conducted in randomly selected half of the counties/cities in 22 provinces and municipalities directly under the Central Government of China. This chapter focuses on studying older adults aged 85 to 105 since prior literature considered persons who reported age 106 or higher as invalid cases (Zeng et al., 2002).

8.2.2 Measures

As stated in Chap. 7, health lifestyle measures used in this study are largely consistent with those used in previous analyses. The measure of leisure activities relies on the survey question asking whether the respondent participated in any leisure activities on a daily basis. The activities included: (1) reading newspapers/books, (2) raising domestic animals, (3) playing cards and/or mah-jong, (4) watching television and/or listening to radio, and (5) gardening work. This study generates a dichotomous variable which is coded as "1" if the respondent reported that he or she participated in at least one of the above five leisure activities almost every day. The variable is coded as "0" if otherwise.

Health outcomes are measured in the same way as they are measured in Chap. 4. In this part of the analysis, demographic and socioeconomic (SES) variables are also included but they are treated as control variables. Detailed description of these variables can be found in Chap. 4.

8.2.3 Methods

This part of the analysis also considers multiple waves of CLHLS data as repeated cross-section surveys. A cross-classified random-effects model (CCREM) is used to take into account the embeddedness of respondents in CLHLS surveys within a time period and a birth cohort (Yang & Land, 2006). Specifically, the model estimated fixed effects of individual-level covariates and random effects of period (or time) and birth cohort. Individuals are the level-1 unit of analysis. Periods (or time) and birth cohort become the level-2 units of analysis. In the analysis of the CLHLS data, an individual respondent is nested in and cross-classified by a higher level context, which is period (or time) and cohort.

Beyond providing random period effects, CCREMs allow for random effects of independent variables, which is used to test for cross-period and cross-cohort variations in the effects of key independent variable, i.e. leisure activity variable. Period-specific random-effects and cohort-specific random-effects coefficients suggest potential across-period and across-cohort changes in the effects of frequently engaging in leisure activities on the oldest old's health status, respectively. This study specifies models of the oldest old's health status as a quadratic function of age.

Several models are presented in the analysis. The first model only includes age, age-squared and the control variables as fixed effects. This model measures the overall effects of period on the oldest old's health status, controlling for age and demographic and socioeconomic variables. The following models include the leisure activity measure, controlling for the demographic and socioeconomic factors. The full individual level or level-1 equation can be expressed as:

$$\mathrm{Y}_{\mathrm{ijk}} = \beta_{0\mathrm{jk}} + \beta_{1\mathrm{jk}}\mathrm{AGE} + \beta_{2\mathrm{jk}}\mathrm{AGE}^2 + \beta_{3\mathrm{jk}}\mathrm{X} + \sum_{p=10}^{p} \beta_{\mathrm{p}}\mathrm{Xp} + \mathrm{e}_{\mathrm{ijk}} \tag{8.1}$$

where Y_{ijk} represents health status of the *i*th respondent for i = 1, …, n*j* individuals within the *j*th period for j = 1, …. β_{0j} is the intercept or for respondents in period *j*; "AGE" and "AGE^2" symbolize age and age-squared, respectively. "X" indicates independent variables; i.e., the leisure activity variable. β_p represents other individual-level fixed effects where *k* is the maximum number of covariates. e_{ij} represents an individual level random error term.

The level-2 model can be expressed as:

$$\beta_{0j} = \gamma_0 + p_{0j} + c_{0k} \tag{8.2}$$

This model specifies the overall mean vary from period to period and from cohort to cohort. γ_0 is the model intercept, which is the expected mean at the zero value of all level-1 variables averaged over all periods and cohorts. Here p_{0j} and c_{0k} are the residual random effects of period and cohort, respectively. The cell mean, $\beta_{0j,}$ is equal to the sum of the overall mean or intercept (γ_0), and the residual random effect of period j (p_{0j}) and the residual random effect of cohort (c_{0k}). The residual random effects of period j (p_{0j}) and cohort k (c_{0k}) examine the effects of each period and cohort on a specific health measure. When adding (8.2) to (8.1), it yields the level-1 and level-2 combined model:

$$Y_{ijk} = \gamma_0 + \beta_{0jk} + \beta_{1jk}\text{AGE} + \beta_{2jk}\text{AGE}^2 + \beta_{3jk}\text{X} + p_{0j} + c_{0k} + e_{ijk} \tag{8.3}$$

where, within each survey year j, respondent i's health is modeled as a function of his or her age, age-squared, the leisure activity variable and other control variables. In the above model, γ_0 represents the fixed-effects coefficients for the leisure activity variable; the symbols of p_{0j} and c_{0k} represent the period-specific and cohort-specific effects of the leisure activity variable, respectively. The model tests for random effects of leisure activity patterns on the respondent's health across periods and cohorts. All continuous variables are sorted and centered around their overall mean.

Since there are eight waves of survey data, survey years for the eight waves are considered as the eight periods. I have developed 13 birth cohorts in the analyses (pre-1897, 1898–1900, 1901–03, 1904–06, 1907–09, 1910–12, 1913–15, 1916–18, 1919–21, 1922–24, 1925–27, 1928–30 and 1931–33). The 1934 and later birth cohorts are dropped from the analyses since those cohorts do not fall into the age range of 85 to 105. When showing descriptive results, all 13 birth cohorts are shown in graphs. When conducting regression analyses, the 1928–30 and 1931–33 birth cohorts are combined into one birth cohort considering the number of respondents who were born in and after 1928 is limited.

8.3 Results

8.3.1 Descriptive Statistics

The chapter first describes the changing patterns of the oldest old's leisure activity patterns when time period and birth cohorts are considered. The study relies on graphs to show how leisure activity patterns among Chinese oldest old have changed by time period and by birth cohort. Figure 8.1 shows percentage distributions of the oldest old who engaged in leisure activities both in the past and at present during the studied period. The results show that such percentages range from 15.5% to 61.6%, suggesting that there are higher percentages of younger birth cohorts who participated in leisure activities as compared to older birth cohorts. Similar to physical activities, percentages of the oldest old who participated in leisure activities show a declining trend. But the declining trend is not as deep as physical activities.

8.3.2 Regression Results Based on APC Analyses

Table 8.1 presents regression results from the CCREMs when including control variables, i.e., demographic and socioeconomic factors, as well as the physical and leisure activity variables in the analyses. The effects of demographic and SES variables have been stated in Chap. 7, which will not be described here. For the fixed

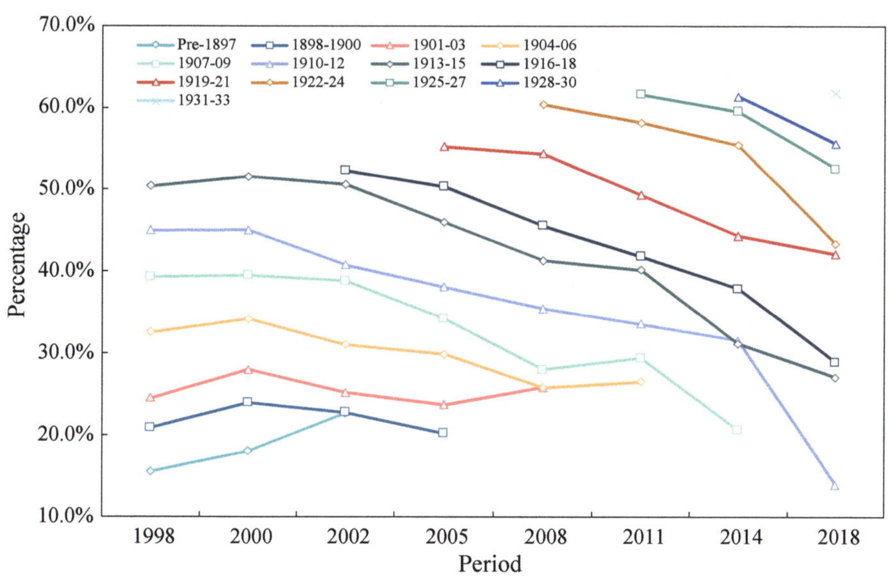

Fig. 8.1 Percentage of respondents who participated in leisure activities both in the past and at present (Unweighted). (Sources: 1998–2018 Chinese Longitudinal Healthy Longevity Survey (CLHLS))

Table 8.1 Regression results of health outcome measures on leisure activity and other control variables from CCREMs: Chinese aged 85–105

Variables	Self-rated health		Chronic disease		Cognitive function		Positive feeling		Negative feeling	
	Coef.	P Value	Coef.	P Value	Coef.	P Value	Coef.	P Value	Coef.	P Value
Intercept	3.212	<0.001	0.870	<0.001	18.389	<0.001	9.591	<0.001	7.910	<0.001
Control variables										
Age	0.007	<0.001	−0.013	<0.001	−0.334	<0.001	−0.002	0.443	−0.016	<0.001
Age^2	0.000	0.147	0.000	0.288	0.001	0.281	−0.001	0.111	0.000	0.256
Sex (ref. = female)	0.032	<0.001	−0.077	<0.001	1.413	<0.001	0.013	0.602	−0.158	<0.001
Residence (ref. = rural)	0.000	0.956	0.282	<0.001	−0.030	<0.001	0.215	<0.001	−0.185	<0.001
Being married	−0.066	<0.001	0.072	<0.001	0.425	<0.001	−0.007	0.821	−0.380	<0.001
Schooling years	−0.002	0.162	0.047	<0.001	0.159	<0.001	0.033	<0.001	−0.021	<0.001
Often going to bed hungry as a child	0.043	<0.001	0.019	0.125	0.530	<0.001	0.160	<0.001	−0.196	<0.001
Activity variables										
Physical activity	0.303	<0.001	−0.050	0.001	1.526	<0.001	0.753	<0.001	−0.568	<0.001
Leisure activity	0.277	<0.001	−0.025	0.557	3.125	<0.001	0.830	<0.001	−0.639	<0.001
Random effect component										
Period										
Intercept	0.010	0.034	0.049	0.034	0.305	0.038	0.047	0.034	0.096	0.033
Leisure activity	0.001	0.121	0.011	0.045	0.446	0.061	0.001	0.374	0.025	0.051
Cohort										
Intercept	0.000	0.132	0.006	0.020	0.113	0.032	0.002	0.156	0.002	0.129
Leisure activity	0	–	0	–	1.061	0.023	0	–	0	–
AIC	122,298.6		167,891.4		302,853.4		213,242.8		201,545.7	
BIC	122,290.6		167,883.4		302,843.4		213,234.8		201,537.7	

Source: 1998–2018 Chinese Longitudinal Healthy Longevity Survey (CLHLS)

effects in level 1, results show that the oldest old who participated in leisure activities both in the past and at present also show better SRH, cognitive function, subjective wellbeing and they have a fewer number of chronic diseases. These results are largely consistent with findings presented in previous studies.

The following parts present significant period and cohort effects in the association between leisure activities and Chinese oldest old's health measures. Figure 8.2 shows that the oldest old who engaged in at least one leisure activity almost every day have better predicted SRH than the reference group that did not engage in any leisure activities frequently. The period effects are shown in a general declining trend in predicted SRH over time with a slight increase in years 2008 and 2018. The results do not show significant cohort effects.

Significant period effects are also observed when using leisure activities to predict the number of chronic diseases. As Fig. 8.3 shows, the less active group shows a fewer number of chronic diseases before year 2008. But after 2008, crossovers occur and the more active group shows a greater number of chronic diseases in more recent years. One possible explanation of the phenomenon relates to the oldest old's SES. Those who participated in leisure activities more frequently could also be the ones that are socioeconomically doing better. Year 2008 can be considered as a threshold when China experienced an economic hardship and the economy gradually recovered afterwards. The SES differentials among various social groups enlarged afterwards. The enlarged SES differences may have led to a rising gap in the number of chronic diseases.

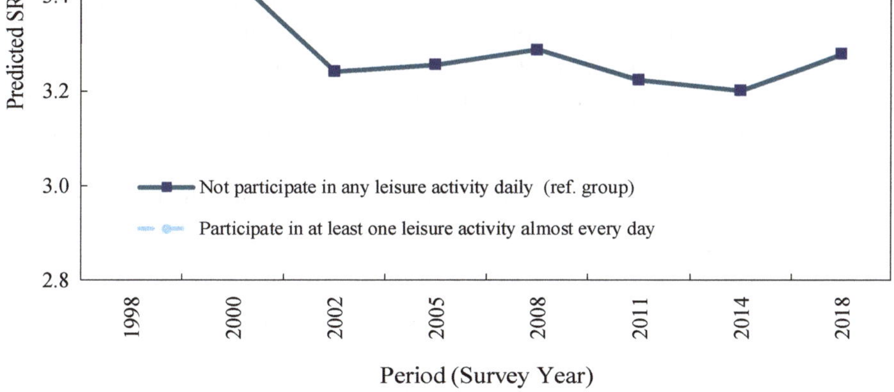

Fig. 8.2 Predicted self-rated health by leisure activity over time period. (Sources: 1998–2018 Chinese Longitudinal Healthy Longevity Survey (CLHLS))

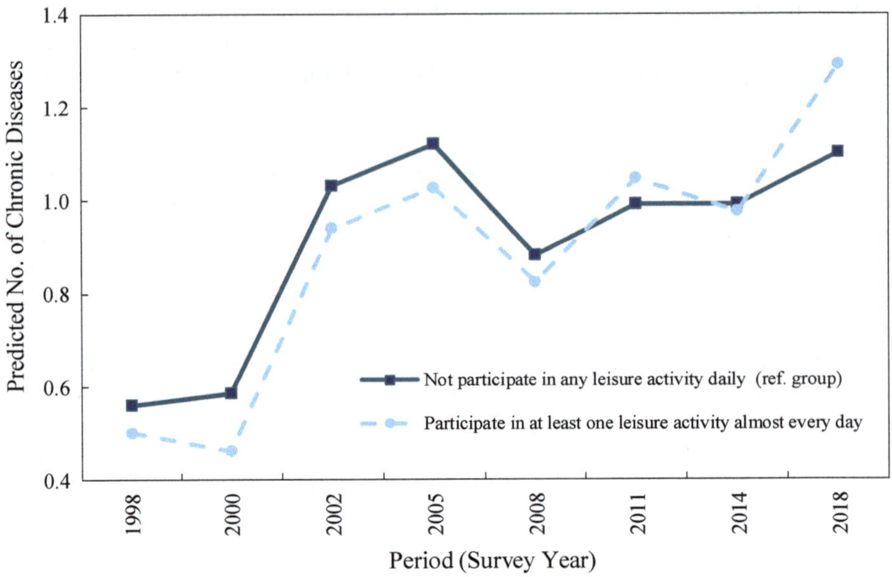

Fig. 8.3 Predicted number of chronic diseases by leisure activity over time period. (Sources: 1998–2018 Chinese Longitudinal Healthy Longevity Survey (CLHLS))

Figures 8.4a and 8.4b show significant period and cohort effects when using leisure activities to predict cognitive function scores, respectively. As it is shown by both figures, those who engaged in leisure activities more frequently tend to have better cognitive function scores than those who are less active. Figure 8.4a indicates that overall, there is a slight increasing trend in predicted cognitive function scores for both groups during the studied time frame. But the predicted cognitive function scores show a more apparent increasing trend over time for the more active group.

As to the cohort effect, Fig. 8.4b shows that the cognitive function stays relatively stable across birth cohorts for the less active group. But the cognitive function for the group that participated in at least one leisure activity daily shows a declining trend across birth cohorts. The results may suggest that the positive effect of participating in leisure activities on maintaining cognitive function has been decreasing in more recent birth cohorts.

Significant period effects are also found when using the leisure activity variable to predict the oldest old's subjective wellbeing measures, i.e. positive feeling and negative feeling scores. No significant cohort effects are observed here. Figures 8.5 and 8.6 demonstrate predicted positive and negative feeling scores by time period, respectively. According to both figures, frequent leisure activity participants show better subjective wellbeing than those who did not engage in any leisure activity daily. Similar to previous findings, year 2008 is the year that the respondents reported the lowest positive feeling scores. The negative feeling scores show a general declining trend over time, suggesting that Chinese oldest old's subjective wellbeing is getting better over time.

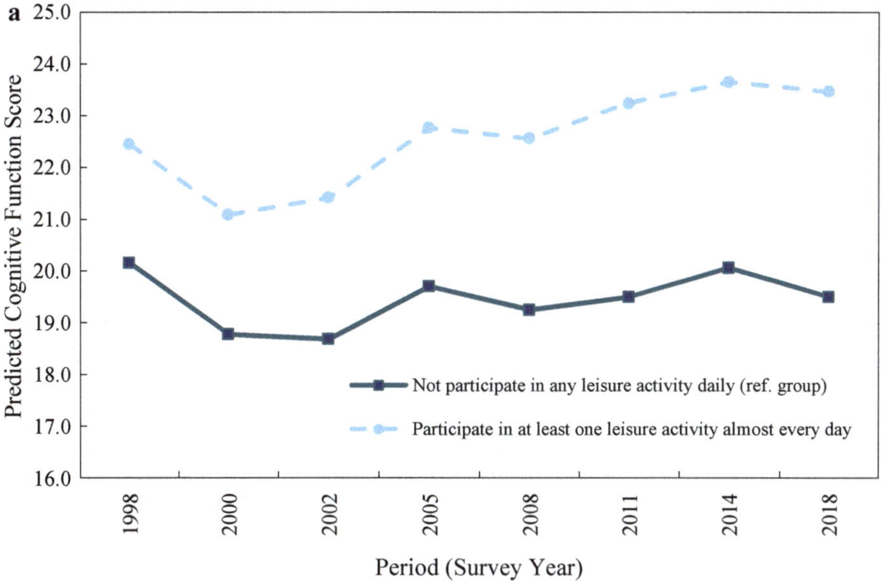

Fig. 8.4a Predicted number of cognitive function score by leisure activity over time period. (Sources: 1998–2018 Chinese Longitudinal Healthy Longevity Survey (CLHLS))

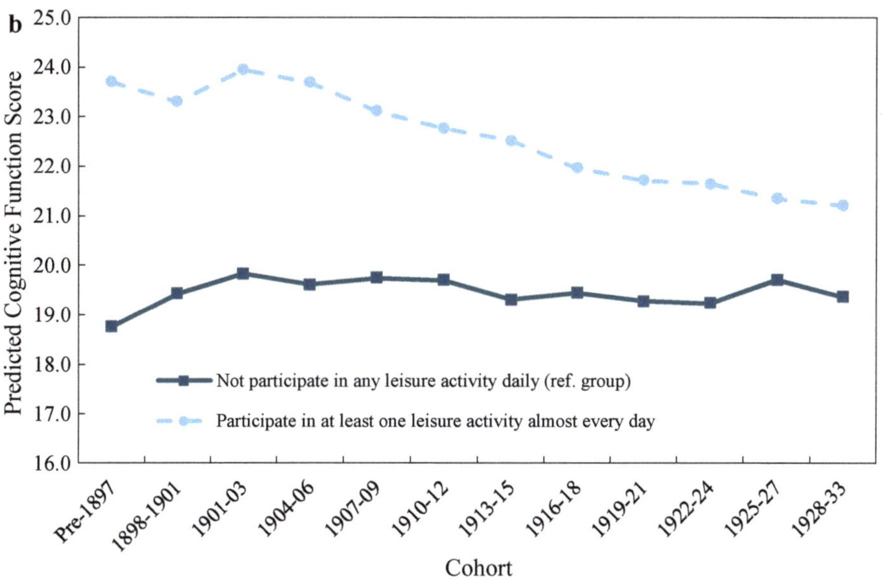

Fig. 8.4b Predicted number of cognitive function score by leisure activity over birth cohort. (Sources: 1998–2018 Chinese Longitudinal Healthy Longevity Survey (CLHLS))

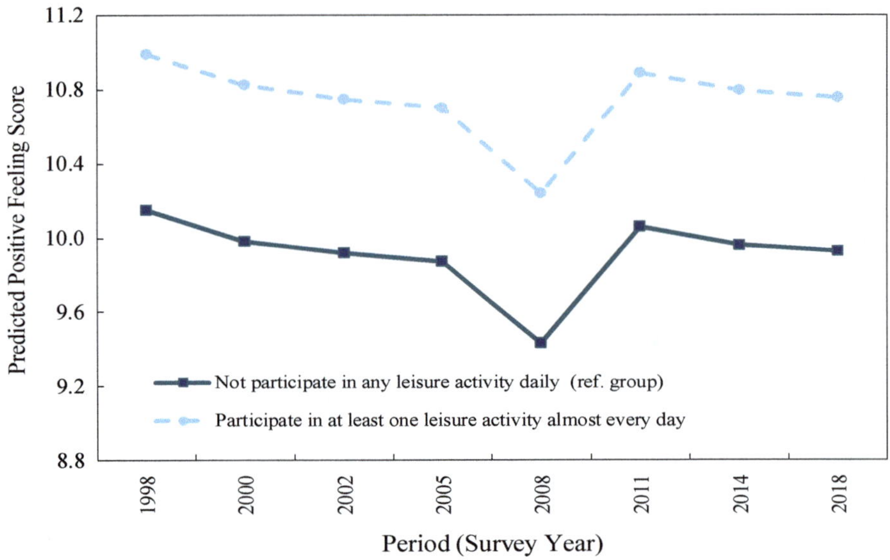

Fig. 8.5 Predicted positive feeling score by leisure activity over time period. (Sources: 1998–2018 Chinese Longitudinal Healthy Longevity Survey (CLHLS))

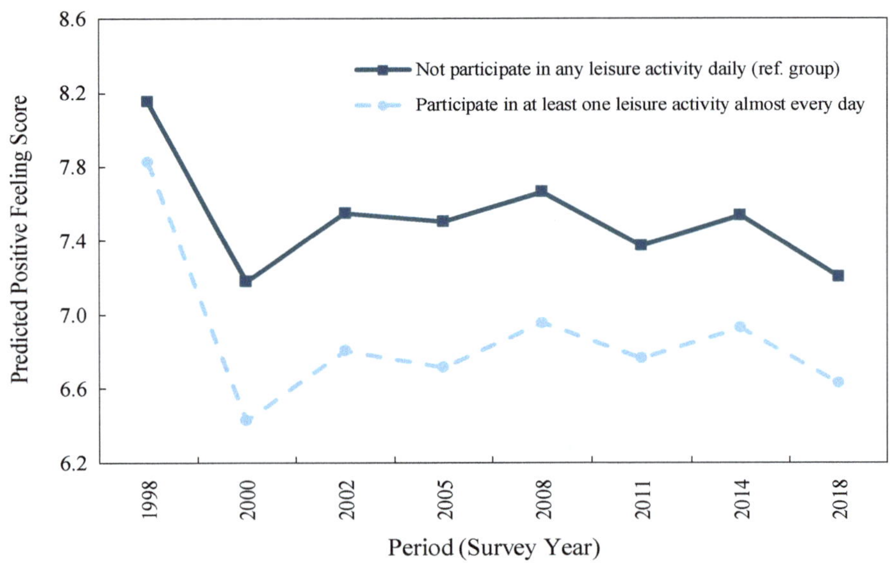

Fig. 8.6 Predicted negative feeling score by leisure activity over time period. (Sources: 1998–2018 Chinese Longitudinal Healthy Longevity Survey (CLHLS))

8.4 Conclusion and Discussion

This chapter has concentrated on exploring the associations between Chinese oldest old's health and their leisure activity patterns. Even with abundant literature on leisure activity participation and older adults' health, our knowledge is very limited in terms of how period and cohort show compound effects on the oldest old's health due to their leisure activity patterns. This chapter intends to improve previous literature by studying the changing patterns of Chinese oldest old's leisure activity patterns over time and across birth cohorts. Some new findings are drawn from constructing CCREMs by using the time-series data on Chinese oldest old's health that span 20 years.

First of all, the results show period and cohort variations in Chinese oldest old's leisure activity patterns during the studied 20-year span. The descriptive results show that the percentage of the oldest old who engaged in at least one leisure activity daily has been declining over time. This can be explained by the fact the respondents were getting older over time and they became less active in leisure activities as compared to previous years. But the cohort differentials are more apparent when it comes to the leisure activity measure. Such findings suggest that younger birth cohorts become more active, especially in leisure activities.

Moreover, the health disparities predicted by the oldest old's leisure activity patterns vary by period and cohort. The results imply that those who engaged in at least one leisure activity almost every day show better SRH, cognitive function and subjective wellbeing. But they tend to have a greater number of chronic diseases than those who did not engage in any leisure activities on a daily basis. Over time, when using the leisure activity variable to predict the oldest old's health status, the predicted SRH shows a declining trend, whereas the cognitive function is getting better over time. A rising trend is observed in the predicted number of chronic diseases. And a V-shape is observed in positive feeling scores due to negative social and economic events in year 2008.

In conclusion, this chapter has attempted to investigate the period and cohort effects in the relationship between Chinese oldest old's leisure activity patterns and their health status. This research showed that younger birth cohorts are more active in participating leisure activities. The health disparities between the sedentary and more active groups in leisure activities revealed by this study have again emphasized the importance of participating in social activities in maintaining older adults' healthy status. Participating in leisure activities help older adults to remain active, which prevents them from degeneration.

When using leisure activity patterns to predict Chinese oldest old's health status, significant period and cohort effects are shown with period effects being more prominent than cohort effects. These findings contribute to the existing analyses and theories on health lifestyles and older adults' health. Even with the contribution of this analysis, the social, economic, political and cultural factors that may have caused period and cohort variations have not been investigated by this research.

Future research may consider exploring factors behind the period and cohort changes to offer a more comprehensive understanding of the oldest old's health and their leisure activity patterns.

References

Adjei, N. K., & Brand, T. (2018). Investigating the associations between productive housework activities, sleep hours and self-reported health among elderly men and women in western industrialised countries. *BMC Public Health, 18*, 110.

Araújo, J., Ramos, E., & Lopes, C. (2011). Lifestyles and self-rated health, in Portuguese elderly from rural and urban area (in Portuguese). *Acta Médica Portuguesa, 24*, 79–88.

Atchley, R. C. (1999). *Continuity and adaptation in aging: Creating positive experiences*. Johns Hopkins University Press.

Bae, S. M. (2020). The association between health-related factors, physical and mental diseases, social activities, and cognitive function in elderly Koreans: A population-based cross-sectional study. *Psychogeriatrics, 20*, 654–662.

Bhattacharya, P. K., Deka, K., Barman, B., & Jamil, M. (2023). Association between physical fitness and perceived Well-being in functionally independent community dwelling elderly of north-eastern India. *Acta medica Lituanica, 30*(1), 6–18.

Bum, C.-H., Johnson, J. A., & Choi, C. (2020). Healthy aging and happiness in the Korean elderly based upon leisure activity type. *Iranian Journal of Public Health, 49*, 454–462.

Chang, H., Huang, Y., & Liu, X. (2022). Loneliness status and related factors among the Hakka elderly in Fujian, China: Based on the health ecological model. *Journal of Environmental and Public Health*, 2633297.

Confortin, S. C., Giehl, M. W., Antes, D. L., Schneider, I. J., & d'Orsi, E. (2015). Positive self-rated health in the elderly: A population-based study in the south of Brazil. *Cadernos de Saúde Pública, 31*, 1049–1060.

Du, L. (2022). Study on the influence of different types of social participation on the physical and mental health of the elderly (in Chinese). *Advances in Applied Mathematics, 11*, 3883–3890.

Fan, Y., Beilei, W., & Fubaihui, W. (2023). Influence of physical exercise on depression of Chinese elderly (in Chinese). *China Sport Science and Technology, 59*, 38–43.

Hansdottir, H., Jonsdottir, M. K., Fisher, D. E., Eiriksdottir, G., Jonsson, P. V., & Gudnason, V. (2022). Creativity, leisure activities, social engagement and cognitive impairment: The AGES-Reykjavik study. *Aging Clinical and Experimental Research, 34*, 1027–1035.

Jung, M. S., Kim, H., Lee, Y., Kim, M., & Chung, E. (2017). Different effects of cognitive and non-exercise physical leisure activities on cognitive function by age in elderly Korean individuals. *Osong Public Health and Research Perspectives, 8*, 308–317.

Komatsu, M., Obayashi, K., Tomioka, K., Morikawa, M., Jojima, N., Okamoto, N., Kurumatani, N., & Saeki, K. (2019). The interaction effect between physical and cultural leisure activities on the subsequent decline of instrumental ADL: The Fujiwara-kyo study. *Environmental Health and Preventive Medicine, 24*, 71.

Li, Z., & Zhenhui, C. (2017). The impact of leisure activities on Chinese elderly's cognitive function (in Chinese). *Journal of Psychologyical Science, 40*, 380–387.

Li, J., Hsu, C. C., & Lin, C. T. (2019). Leisure participation behavior and psychological well-being of elderly adults: An empirical study of Tai Chi Chuan in China. *International Journal of Environmental Research and Public Health, 16*(18), 3387.

Li, W., Sun, H., Xu, W., Ma, W., Yuan, X., Wu, H., & Kou, C. (2021). Leisure activity and cognitive function among Chinese old adults: The multiple mediation effect of anxiety and loneliness. *Journal of Affective Disorders, 294*, 137–142.

Li, X., Zhang, J., Hou, R., Zheng, M., Singh, M., Li, H., Li, C., Zhang, X., Yang, X., Wu, L., Wang, Y., & Zheng, D. (2022). Bidirectional associations of intellectual and social activities with cognitive function among middle-aged and elderly adults in China. *Journal of Affective Disorders, 319*, 83–89.

Michele, J., Guillaume, M., Alain, T., Nathalie, B., Claude, F., & Kamel, G. (2019). Social and leisure activity profiles and well-being among the older adults: A longitudinal study. *Aging & Mental Health, 23*, 77–83.

Nimrod, G., & Kleiber, D. A. (2007). Reconsidering change and continuity in later life: Toward an innovation theory of successful aging. *International Journal of Aging Human Development, 65*, 1–22.

Silva, D. A. S. (2021). Longer leisure walking time is associated with positive self-rated health among adults and older adults: A Brazilian nationwide study. *Peer J, 9*, e11471.

Xiaoyun, C., Yuhang, W., Yu, C., Wei, Z., Songbo, H., & Huilie, Z. (2022). Correlation between leisure activities and activities of daily living in the elderly: A prospective cohort study of 80-year-old individuals (in Chinese). *Journal of China Medical University, 51*, 1079–1089.

Xu, X. Y., Wang, S. S., Niu, L., Leung, I. S. H., & Tian, Q. B. (2022). Association of leisure activity changes and reversion from mild cognitive impairment to normal cognitive function among older adults: A prospective cohort study. *Frontiers in Public Health, 10*, 1035762.

Yang, Y., & Land, K. C. (2006). A mixed models approach to the age-period-cohort analysis of repeated cross-section surveys, with an application to data on trends in verbal test scores. *Sociological Methodology, 36*, 75–97.

Zeng, Y., Vaupel, J., Xiao, Z., Zhang, C., & Liu, Y. (2002). Sociodemographic and health profiles of the oldest old in China. *Population and Development Review, 28*, 251–273.

Zhang, K., & Zhang, W. (2021). Leisure activities and healthy aging. In D. Gu & D. M. Dupre (Eds.), *Encyclopedia of gerontology and population aging* (pp. 2884–2888). Springer Publisher.

Zhang, W., Feng, Q., Lacanienta, J., & Zhen, Z. (2017). Leisure participation and subjective Well-being: Exploring gender differences among elderly in Shanghai, China. *Archives of Gerontology and Geriatrics, 69*, 45–54.

Zhang, W., Zhang, Y., & Wang, J. (2022). Effect and mechanism of reading habits on physical and mental health among the elderly: Evidence from China. *Frontiers in Public Health, 10*, 1031939.

Zhao, Y. (2022). Study on the relationship of leisure sports tourism with the health of the elderly. *Revista Brasileira de Medicina do Esporte, 28*, 432–435.

Zhu, X., Qiu, C., Zeng, Y., & Li, J. (2017). Leisure activities, education, and cognitive impairment in Chinese older adults: A population-based longitudinal study. *International Psychogeriatrics, 29*, 727–739.

Smoking, Alcohol Consumption and the Oldest Old's Health

9

Personal lifestyle factors have long been found to be influential on morbidity and mortality rates of a population (Pan et al., 2020). These factors often include physical activities, smoking, drinking, diet, et al. Scholars indicated that that a healthier lifestyle could reduce morbidity and prolong life expectancy (Almeida De et al., 2020; Araújo et al., 2011). Thus, adopting a healthy lifestyle to prolong life expectancy has also entered the public agenda to promote healthy aging. Smoking and drinking are considered as important dimensions of lifestyles, and both of those behaviors have been documented to jeopardize individual health. This chapter pays a particular attention to Chinese oldest old's smoking and drinking behaviors and takes a longitudinal point of view to explore how smoking and drinking behaviors among Chinese oldest old have changed by time period and across birth cohorts. In addition, it answers the question that how period and cohort show complex effects in the association between alcohol consumption, smoking and Chinese oldest old's health.

9.1 Theories Linking Drinking, Smoking and Individual Health

When explaining the negative effects of drinking and smoking on individual health, several theories have been proposed to elucidate the risks of drinking and smoking behaviors. This section reviews these theories and provides theoretical basis for empirical analyses conducted in the following sections. The first theory is called *toxicity theory*. This theory argued that smoking and drinking bring harmful substances to the body, which is detrimental to human tissues and organs (Hecht, 2006). For example, smoking produces nicotine, carbon monoxide and carcinogens, all of which are risk factors that may cause lung cancers (Ambati, 2021; Hecht, 2006). Alcohol is also a toxic chemical substance. Consuming alcohol excessively can

L. Zhang, *Living Longer and Healthier at Older Ages*, Advances in Studies of Aging and Health 3, https://doi.org/10.1007/978-3-031-69773-9_9

damage the brain and nerves, which causes mental confusion and cognitive impairment (Panza et al., 2009; Zhang et al., 2020).

The second theory is *metabolic disorder theory*. This theory contended that as people age, their metabolic functions naturally decline. Excessive smoking and drinking can further exacerbate the decline in metabolic function among older adults and affect their absorption, utilization and metabolism of nutrients (Behl et al., 2023; Harris et al., 2016).

Immune system suppression theory is the third theory that has been commonly used to explain damages of smoking and drinking bringing to human beings. The theory argued that smoking and excessive drinking have a significant destructive effect on human immune function (Sarkar et al., 2015), especially for older adults who may already have weakened immune systems (Aspinall & Lang, 2018; Sarkar et al., 2015). As the immune function declines in older adults, their risks of developing diseases increases (Aspinall & Lang, 2018). Long-term smoking and drinking can lead to insulin resistance, diabetes, obesity and other health related issues, which increases the likelihood of developing various chronic diseases (Ng et al., 2020; Wang et al., 2019).

The fourth theory is the *inflammation theory*. The theory considered smoking and drinking as risk factors that can cause chronic inflammation (Attard et al., 2017). This is because smoking and drinking can produce many harmful substances (Sanada et al., 2018) and these substances can trigger inflammation and damage cells and tissues of human beings (Traber et al., 2000). Inflammation may further lead to a range of chronic diseases such as cardiovascular diseases, liver diseases and cancer (Sanada et al., 2018; Traber et al., 2000).

The next theory is derived from the *health demand model*. The theory proposed that the stock of health capital determines the number of healthy days that people have. Excessive smoking and drinking reduce people's health stock (Lahiri & Li, 2020). For older adults, in particular, as they age, the overall health stock tends to decline. The continuity of smoking and drinking accelerates the depletion of people's health capital. Research found that older adults who quit smoking actually enjoy longer lives as compared to those who continued smoking (Taylor Jr et al., 2002). Further studies have shown that quitting smoking also increased individual lifespans by reducing smoking-related diseases (Ostbye & Taylor, 2004).

The last theory is called *"Diet-Smoking-Gender" interaction theory*. The theory contended that the "Diet-Smoking-Gender" three-way interaction has significant effects on cognitive impairment among older adults. Researchers have documented a correlation between higher dietary diversity and lower cognitive dysfunction among older adults (Chen et al., 2022; Huang et al., 2021; Otsuka et al., 2017). Meanwhile, researchers indicated that smoking behavior negatively influenced the above relationship as smoking behavior counteracted the probability of lower cognitive impairment that is associated with a higher dietary diversity among older adults (Chen et al., 2022). In sum, the reviewed theories have explained the association between smoking and drinking and individual health from a variety of perspectives. They provide a theoretical basis for studying the influence of smoking on older adults' health.

9.2 Empirical Research on Smoking and Older Adults' Health

9.2.1 Cross-sectional Studies

Smoking is recognized as a significant risk factor for global population health. Over five million deaths are caused by smoking annually worldwide (World Health Organization, 2021). Numerous studies have consistently demonstrated the detrimental effects of smoking on individual health (Escobedo et al., 1997; Yang & Zikos, 2023). This section first reviews empirical findings from a cross-sectional point of view and it then moves to longitudinal analyses on this subject. Cross-sectional findings on the influence of smoking on older adults' health centered on the following areas:

Prior studies showed that smoking increased the risks of developing multiple types of chronic diseases, such as lung cancer, cardiovascular diseases and respiratory disorders (Jing et al., 2021; Strine et al., 2005). Globally, lung cancer is the second leading cause of death, and its incidence is closely associated with smoking (Domingo-Relloso et al., 2023). A study revealed that the longer the duration of smoking and the greater amount of cigarette an individual smoked, the higher the risk of developing lung cancer diseases (Sarrio et al., 2010). This is because smoking can weaken people's immune system, particularly for older adults, and thereby largely increase the risks of having chronic illnesses (Agnihotri & Gaur, 2014). Due to vascular aging and related functional decline, older adults are more vulnerable to the detrimental effects of smoking, which eventually leads to a range of cardiovascular diseases, hypertension, myocardial infarction stroke and so forth (Chen et al., 2015).

The second area of related studies touched on disease incidence and mortality rates. Reports have indicated that smoking is associated with higher disease incidence and mortality rates. Individuals aged 60 and above accounted for approximately 70.0% of the affected population (Burns, 2000).

The third area of prior studies dealt with oral health related issues. Researchers considered smoking as a risk factor for poor oral health among older adults. Multiple studies have provided substantial evidence of a significant correlation between smoking and worse oral health among elders (Agnihotri & Gaur, 2014). Researchers found that smoking can lead to adverse oral health outcomes, including widespread tooth loss, periodontal diseases and oral cancer (Jeong et al., 2022). The presence of oral diseases eventually led to detrimental dietary habits and weight loss among older adults (Lesnoff-Caravaglia, 2007).

Along with the aging process, older adults may also experience osteoporosis and fractures. Smoking is considered as a significant risk factor for osteoporosis in older ages, as it leads to a decrease in bone density and an increased risk of fractures (Hou et al., 2023; Tamaki et al., 2011). Furthermore, the relationship between smoking and cognitive function among older adults has also been repeatedly documented by the existing literature. Empirical studies have demonstrated that smoking accelerated cognitive decline among older adults and increased the risks of developing Alzheimer's diseases (Mons et al., 2013).

9.2.2 Longitudinal Studies

As compared to the abundant literature that examined smoking and older adults' health from a cross-sectional point of view, longitudinal analyses seem to be scarce. The longitudinal studies documenting the time effects linking smoking and older adults' health can be categorized into several dimensions. The first dimension touched on smoking and older adults' mortality. Through conducting long-term follow-up studies, researchers found that elder smokers had a significantly higher mortality rate as compared to non-smokers (Kihara et al., 2022; Lim et al., 2013; Lim et al., 2012; Nash et al., 2017). Furthermore, prior studies indicated that there was no significant difference in mortality rates between male and female elder smokers over time (Lam et al., 2015). In addition, a predictive study on the relationship between smoking and lifespan among older adults revealed that smoking continued to be a significant contributor to the mortality of older Americans aged 55 to 85 from 2011 to 2040 (Cao, 2016).

The second dimension focused on investigating smoking and the risks of developing smoking-related diseases among older adults. The results of these longitudinal studies have shown that elder smokers had a higher risk of developing smoking-rated diseases such as coronary heart disease, chronic kidney disease, dementia and cancer as compared to non-smokers (Barengo et al., 2019; Lu et al., 2020; Tolstrup et al., 2014; Zammit et al., 2016). Furthermore, smoking exacerbated the severity of these diseases among older adults. An analysis of 3-year monitoring data on older patients with ocular diseases revealed that current smokers with cataracts were more likely to experience visual impairment compared to those who had never smoked (Barengo et al., 2019). Among patients who smoked consistently throughout 3 years, their visual impairment was most significant, indicating that smoking worsened cataract-related visual impairment in older patients. However, a follow-up study on 8512 elderly individuals in the UK over 7.8 years found no association between secondhand smoke exposure and myocardial infarction. No differences were observed between non-smokers and ex-smokers (Jefferis et al., 2010).

Besides the time effects, prior studies also linked age to smoking and smoking related diseases. A study conducted in China suggested that when examining the relationship between smoking and mortality, age effects should be considered as there was a significant interaction between age and smoking (Li et al., 2016). Compared to young smokers, older smokers had a higher mortality rate (Li et al., 2016). Furthermore, comparisons among individuals of different age groups have revealed that among current smokers, the younger the age, the greater the risks of developing coronary heart diseases. The older the age, the lower the risks, indicating the presence of age effects (Tolstrup et al., 2010).

9.3 Empirical Research on Alcohol Consumption and Older Adults' Health

9.3.1 Cross-sectional Studies

Alcohol consumption has been documented to have a significant impact on older adults' health. Cross-sectional studies on this subject has touched on the influence of alcohol consumption on various health dimensions among older adults. The first dimension related to chronic diseases as excessive drinking can trigger a range of chronic diseases (Han et al., 2018). For example, studies have found a U-shaped relationship between alcohol intake and the risks of developing diabetes, with excessive drinking increasing such risks among older adults (Shimizu et al., 2014). Prior studies also demonstrated a correlation between heavy alcohol consumption and hypertension among older adults. There was a significantly higher risk of hypertension among heavy drinkers as compared to non-drinkers (Peltzer & Phaswana-Mafuya, 2013).

The second health dimension examined by prior research was osteoporosis and fractures among older adults. Researchers showed a strong correlation between alcohol consumption and osteoporosis and fractures among elders (Mukamal et al., 2007). As alcohol intake increasing, the prevalence of osteoporosis among elders also rose (Lau et al., 2001). Additionally, heavy alcohol consumption was found to be able to damage visceral and nervous systems, which increased the risks of developing Alzheimer's diseases and other types of illnesses. As individuals age, their organ systems and related functions undergo degeneration. Excessive alcohol consumption undoubtedly poses greater harm to visceral and nervous systems (Lewis, 2008). For example, excessive alcohol consumption was found to stimulate the nervous system, leading to neurological impairment and an increased risk of developing mental disorders (Choi & DiNitto, 2021; Saunders et al., 1991). In addition, excessive drinking was found to damage the liver, which increased the risk of developing cirrhosis, cardiovascular diseases and coronary heart disease (Piano et al., 2017).

9.3.2 Longitudinal Studies

In addition to cross-sectional studies, researchers have also conducted longitudinal studies to investigate the relationship between alcohol consumption and older adults' health. Existing literature on the subject mainly focused on studying the risks of developing alcohol-related diseases among older adults. Age effects were found to be significant. Individuals with different levels of lifelong alcohol consumption exhibited different mortality rates. Age was also found to be correlated with disease incidence, with post-World War II cohorts showing higher incidence rates over time (Sacco et al., 2015). A study conducted by scholars at Johns Hopkins University in the United States tracked drinking patterns of individuals aged 55 to 72 on multiple occasions and analyzed the prospective relationship between alcohol consumption and their midlife and late-life cognitive function. Findings showed that older adults who engaged in excessive alcohol consumption during midlife experienced significant impairment in cognitive function in later years (Gross et al., 2011). Another study on older Americans who were surveyed between 1993 and 2000 revealed that as time and age progressing, the incidence rate of alcohol-related diseases surged among older drinkers (Sacco et al., 2015). Higher incidence rates also implied increased mortality rates among older adults. A cohort study on individuals aged 55 to 74 investigated the relationship between alcohol consumption and mortality rates, demonstrating a positive correlation between lifelong alcohol consumption and cancer-related mortality rates (Kunzmann et al., 2018). Additionally, individuals with various levels of lifelong alcohol consumption exhibited different mortality rates with heavy drinkers showing higher mortality rates (Kunzmann et al., 2018).

Other researchers, however, found different results by conducting a longitudinal analysis of the relationship between alcohol consumption and the incidence of central nervous system and cardiovascular diseases in older adults aged 65 and over in Ireland. Researchers compared results from three collected longitudinal data and did not find any changes in risks of developing alcohol-related diseases in older adults over time (Kunzmann et al., 2018). In other words, there was no evidence suggesting that moderate and consistent alcohol consumption among older adults would increase the risk of developing related diseases (Kunzmann et al., 2018). Similarly, a Spanish study on a follow-up cohort of 2,170 older adults aged 60 over during a period of 3.3 years showed a correlation between moderate alcohol consumption and a lower risk of falls in older adults (Ortolá et al., 2017).

To summarize, the existing literature has showed strong evidence that smoking and alcohol consumption have significant influence on seniors' health status although the findings on the association between alcohol consumption and individual health is not conclusive. The majority of these existing studies were cross-sectional, we know very little about the multifaceted effects of age, cohort and period in the association between smoking, alcohol consumption and the oldest old's health. Through analyzing data from CLHLS that extends over 20 years and constructing CCREMs, this chapter intends to explore the changing patterns of Chinese oldest old's smoking and alcohol consumption by time period and across birth cohorts. It also investigates whether significant period and cohort effects exist in the association between the oldest old's health and their smoking and alcohol consumption behaviors.

9.4 Data, Measures and Methods

9.4.1 Data

This chapter analyses data from eight (1998, 2000, 2002, 2005, 2008, 2011–12, 2014 and 2018) waves of the Chinese Longitudinal Healthy Longevity Survey (CLHLS) which was conducted in randomly selected half of the counties/cities in

22 provinces and municipalities directly under the Central Government of China. This chapter focuses on studying older adults aged 85 to 105 since prior literature considered persons who reported age 106 or higher as invalid cases (Zeng et al., 2002).

9.4.2 Measures

Health lifestyle measures used in this study are largely consistent with those used in previous analyses. In the CLHLS, there were no questions asking whether the respondent was a heavy smoker or drinker. But there were questions asking the respondent whether he or she smoked or drank alcohol "in the past" and "at present". The respondent who smoked both in the past and at present is coded as "1" and "0" if otherwise. It is assumed that those individuals who smoked before and were still smoking when the survey was conducted are heavy smokers; the same rationale and coding strategy are also applied to the alcohol consumption variable. The CLHLS questionnaire did have questions asking how much the respondent drank or smoked per day, but the responding rates were less than 20.0% of the total sample. Hence the study does not include any of those variables measuring the respondent's exact amount of cigarette or alcohol consumption due to an extremely large amount of missing values.

Health outcomes are measured in the same way as they are measured in Chap. 4. In this part of the analysis, demographic and socioeconomic (SES) variables are also included but they are treated as control variables. Detailed description of these variables can be found in Chap. 4.

9.4.3 Methods

This analysis considers multiple waves of CLHLS data as repeated cross-section surveys. A cross-classified random-effects model (CCREM) is used to take into account the embeddedness of respondents in CLHLS surveys within a time period and a birth cohort (Yang & Land, 2006). Specifically, the model estimated fixed effects of individual-level covariates and random effects of period (or time) and birth cohort. Individuals are the level-1 unit of analysis. Periods (or time) and cohorts become the level-2 units of analysis. In the analysis of the CLHLS data, an individual respondent is nested in and cross-classified by a higher level context, which is period (or time) and cohort.

Beyond providing random period effects, CCREMs allow for random effects of independent variables, which is used to test for cross-period variations in the effects of key independent variables, i.e. smoking and alcohol use variables. Period-specific random-effects and cohort-specific random-effects coefficients suggest potential across-period and across-cohort changes in the effects of smoking and alcohol use on the oldest old's health status, respectively. This study specifies models of the oldest old's health status as a quadratic function of age.

Several models are presented in the analysis. The first model only includes age, age-squared and the control variables as fixed effects. This model measures the overall effects of period on the oldest old's health status, controlling for age and demographic and socioeconomic variables. The following models include smoking and drinking measures, controlling for demographic and socioeconomic factors. The full individual level or level-1 equation can be expressed as:

$$Y_{ijk} = \beta_{0jk} + \beta_{1jk}AGE + \beta_{2jk}AGE^2 + \beta_{3jk}X + \sum_{p=10}^{P} \beta_p Xp + e_{ijk} \tag{9.1}$$

where Y_{ijk} represents health status of the *i*th respondent for i = 1, …, n*j* individuals within the *j*th period for j = 1, …. β_{0j} is the intercept or for respondents in period *j*; "AGE" and "AGE^2" symbolize age and age-squared, respectively. "X" indicates independent variables; i.e., smoking and drinking variables. β_p represents other individual-level fixed effects where *k* is the maximum number of covariates. e_{ij} represents an individual level random error term.

The level-2 model can be expressed as:

$$\beta_{0j} = \gamma_0 + p_{0j} + c_{0k} \tag{9.2}$$

This model specifies the overall mean vary from period to period and from cohort to cohort. γ_0 is the model intercept, which is the expected mean at the zero value of all level-1 variables averaged over all periods and cohorts. Here p_{0j} and c_{0k} are the residual random effects of period and cohort, respectively. The cell mean, $\beta_{0j,}$ is equal to the sum of the overall mean or intercept (γ_0), and the residual random effect of period j (p_{0j}) and the residual random effect of cohort (c_{0k}). The residual random effects of period j (p_{0j}) and cohort k (c_{0k}) examine the effects of each period and cohort on a specific health measure. When adding (9.2) to (9.1), it yields the level-1 and level-2 combined model:

$$Y_{ijk} = \gamma_0 + \beta_{0jk} + \beta_{1jk}AGE + \beta_{2jk}AGE^2 + \beta_{3jk}X + p_{0j} + c_{0k} + e_{ijk} \tag{9.3}$$

where, within each survey year j, respondent i's health is modeled as a function of his or her age, age-squared, the smoking and drinking variables and other control variables. In the above model, γ_0 represents the fixed-effects coefficients for the smoking and drinking variables; the symbols of p_{0j} and c_{0k} represent the period-specific and cohort-specific effects of the smoking and drinking variables, respectively. The model tests for random effects of smoking and drinking patterns on the respondent's health across periods and cohorts. All continuous variables are sorted and centered around their overall mean.

Since there are eight waves of survey data, survey years for the eight waves are considered as the eight periods. I have developed 13 birth cohorts in the analyses (pre-1897, 1898–1900, 1901–03, 1904–06, 1907–09, 1910–12, 1913–15, 1916–18, 1919–21, 1922–24, 1925–27, 1928–30 and 1931–33). The 1934 and later birth cohorts are dropped from the analyses since those cohorts do not fall into the age range of 85 to 105. When showing descriptive results, all 13 birth cohorts are shown in graphs. When conducting regression analyses, the 1928–30 and 1931–33 birth cohorts are combined into one birth cohort considering the number of respondents who were born in and after 1928 is limited.

9.5 Results

9.5.1 Descriptive Statistics

The descriptive analyses start with presenting changing patterns of the oldest old's smoking and alcohol use behaviors when time period and birth cohorts are considered. Graphs are generated to describe how smoking and alcohol use behaviors among Chinese oldest old have changed by time period and across birth cohorts (see Figs. 9.1a and 9.1b). Figure 9.1a shows that percentages of the oldest old who smoked both in the past and at present change from about 21.6% in 1998 to 1.6% in 2014. A declining trend is shown by time period. In terms of cohort changes, the results show that percentages of constant smokers are getting lower among younger than older birth cohorts.

Figure 9.1b indicates that percentages of respondents who drank alcohol in the past and at present changed from 25.1% in year 1998 to 4.8% in 2014 and 6.6% in 2018. A declining trend is obviously shown in alcohol use over time period. Cohort variations are less as apparent as period changes. Crossovers are observed without showing significant cohort variations. These findings suggest that over time, percentages of constant smokers and alcohol users among the oldest old are decreasing.

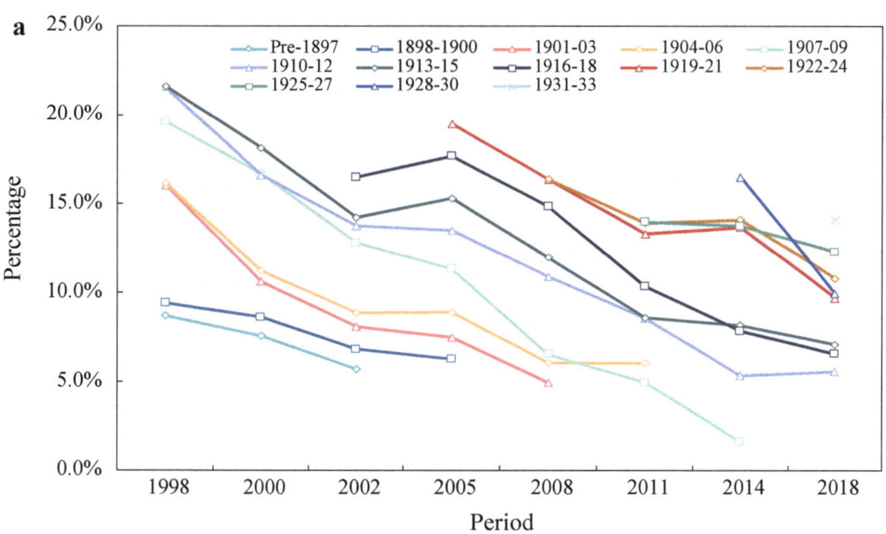

Fig. 9.1a Percentage of respondents who smoked both in the past and at present (Unweighted). (Sources: 1998–2018 Chinese Longitudinal Healthy Longevity Survey (CLHLS))

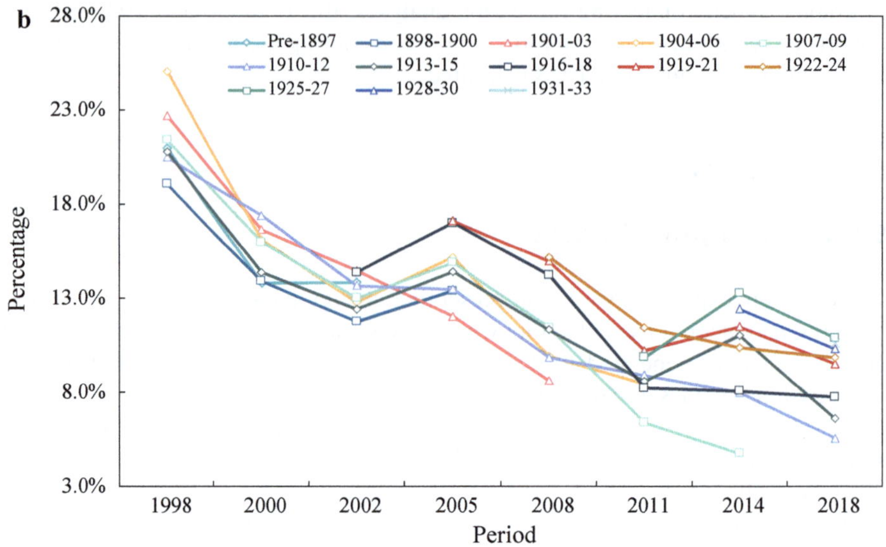

Fig. 9.1b Percentage of respondents who drank alcohol both in the past and at present (Unweighted). (Sources: 1998–2018 Chinese Longitudinal Healthy Longevity Survey (CLHLS))

Also, percentages of constant smokers among Chinese oldest old are getting smaller among younger birth cohorts. But percentages of constant alcohol users do not seem to decline among younger birth cohorts.

9.5.2 Regression Results Based on APC Analyses

Table 9.1 presents regression results from the CCREMs when including control variables, i.e., demographic and socioeconomic factors, as well as the smoking variable in the analyses. The results show that men on average show better SRH and a lower number of chronic diseases; they also have better cognitive function and subjective wellbeing as compared to women. Urban oldest old also show better SRH, cognitive function and subjective wellbeing but a greater number of chronic diseases than their rural counterparts. Married individuals show better cognitive function and subjective wellbeing, but they have worse SRH and more chronic diseases. Higher SES, measured by schooling years and going to bed hungry in childhood, leads to better SRH, cognitive function and subjective wellbeing. But higher SES is also found to be associated with a greater number of chronic diseases.

For the fixed effects in level 1, results show that the smoking variable has significant effects on the first two health measures of the oldest old's health; the alcohol use variables have significant effects on all health measures. Specifically, those who smoked in the past and at present show better SRH and a fewer number of chronic diseases. Alcohol users show better SRH, cognitive function and subjective wellbeing, but a fewer number of chronic diseases. These results are contradictory to

Table 9.1 Regression results of health outcome measures on smoking, alcohol consumption and other control variables from CCREMs: Chinese aged 85–105

	Self-rated health		Chronic disease		Cognitive function		Positive feeling		Negative feeling	
Variables	Coef.	P Value	Coef.	P Value	Coef.	P Value	Coef.	P Value	Coef.	P Value
Intercept	3.289	<0.001	0.903	<0.001	19.228	<0.001	9.824	<0.001	7.728	<0.001
Control variables										
Age	0.002	0.023	−0.015	<0.001	−0.351	<0.001	−0.016	<0.001	−0.005	0.038
Age^2	0.000	0.448	0.000	0.087	0.001	0.402	0.000	0.453	0.000	0.253
Sex (ref. = female)	0.032	0.002	−0.028	0.039	1.557	<0.001	0.046	0.087	−0.166	<0.001
Residence (ref. = rural)	0.055	<0.001	0.268	<0.001	0.426	<0.001	0.362	<0.001	−0.305	<0.001
Being married	−0.040	0.002	0.070	<0.001	0.644	<0.001	0.064	0.056	−0.437	<0.001
Schooling years	0.011	<0.001	0.044	<0.001	0.262	<0.001	0.068	<0.001	−0.049	<0.001
Often going to bed hungry as a child (ref. = yes)	0.051	<0.001	0.015	0.213	0.616	<0.001	0.187	<0.001	−0.216	<0.001
Smoke & alcohol use variables										
Constantly smoking (ref. = yes)	0.060	0.003	−0.139	0.003	0.214	0.164	0.074	0.185	−0.055	0.307
Constantly consuming alcohol (ref. = yes)	0.156	<0.001	−0.196	<0.001	0.756	<0.001	0.313	<0.001	−0.337	<0.001
Random effect component										
Period										
Intercept	0.008	0.035	0.062	0.033	0.414	0.037	0.042	0.034	0.130	0.032
Constantly smoking	0	–	0.005	0.125	0	–	0.010	0.150	0.006	0.221
Constantly consuming alcohol	0.001	0.240	0.002	0.153	0.023	0.331	0	–	0.011	0.138
Cohort										
Intercept	0.000	0.128	0.008	0.018	0.098	0.029	0.003	0.116	0.001	0.187
Constantly smoking	0	–	0	–	0.095	0.150	0	–	0.005	0.236
Constantly consuming alcohol	0.000	0.412	0	–	0.007	0.460	0	–	0	–
AIC	124,230.6		167,836.9		305,736.8		215,545.3		202,979.1	
BIC	124,220.6		167,826.9		305,724.8		215,537.3		202,967.1	

Source: 1998–2018 Chinese Longitudinal Healthy Longevity Survey (CLHLS)

findings presented in previous studies. The chapter discusses the possible reasons that have caused the inconsistency in the conclusion and discussion section.

9.5.2.1 Period and Cohort Effects Predicted by Smoking Variable

The section focuses on showing the random effects which demonstrate significant period and cohort changes in smoking and alcohol use behaviors net of the effects of age and other factors. Since smokers are often alcohol consumers and previous studies often examined smoking and alcohol use effects on individual health at the same time, this chapter includes smoking and alcohol use variables in level 2 simultaneously.

The results from CCREMs have classified the oldest old into four groups: (1) not constant smokers or drinkers, (2) constant smokers but not constant drinkers, (3) constant drinkers but not constant smokers, and (4) both constant drinkers and smokers. Here "constant smoker" is defined as those who smoked both in the past and at present and "constant drinker" is defined as those who drank both in the past and at present. It needs to be pointed out that the first group includes but not restricted to those who never drank or smoked. Individuals who smoked before but quit smoking when the survey was conducted and individuals who did not drink but were drinking when he or she was surveyed and other cases of serials all fall into this group.

The study relies on figures to show significant period and cohort effects when linking smoking and alcohol use behaviors to the oldest old's health. Non-significant period and cohort effects are not presented by figures. Figures 9.2a and 9.2b show significant period and cohort effects when using smoking and drinking behaviors to predict SRH, respectively. Figure 9.2a demonstrates that constant drinkers and smokers have the best SRH, followed by constant drinkers but not constant smokers. The SRH of constant smokers but not constant drinkers rank No.3 and the oldest old who are not constant drinkers or smokers report the lowest SRH scores.

Over time, the SRH for all four groups show a general declining trend and then a slight increasing trend is found in year 2018. The significant cohort effects are shown by a flat bell curve in predicted SRH scores. Specifically, the birth cohort in the middle, i.e., the 1910–12 birth cohort, tends to have better predicted SRH than its earlier and later birth cohorts. These findings suggest that SRH can be influenced by the oldest old's lifetime smoking and drinking behaviors and such influence varies by time period and birth cohort.

When using smoking and drinking variables to predict the number of chronic diseases, only significant period effects are shown, which is described by Fig. 9.3. The figure shows that those who are not constant smokers or drinkers tend to have a greater predicted number of chronic diseases, followed by the groups composed by constant smokers but not constant drinkers, and constant drinkers but not constant smokers. The oldest old who are constant smokers and drinkers have the lowest predicted number of chronic diseases.

During 1998 to 2018, the predicted number of chronic diseases for the oldest old show an increasing trend. For instance, such a number change from 0.3 to 0.7 for constant smokers and constant drinkers group during 1998 to 2018. And the values

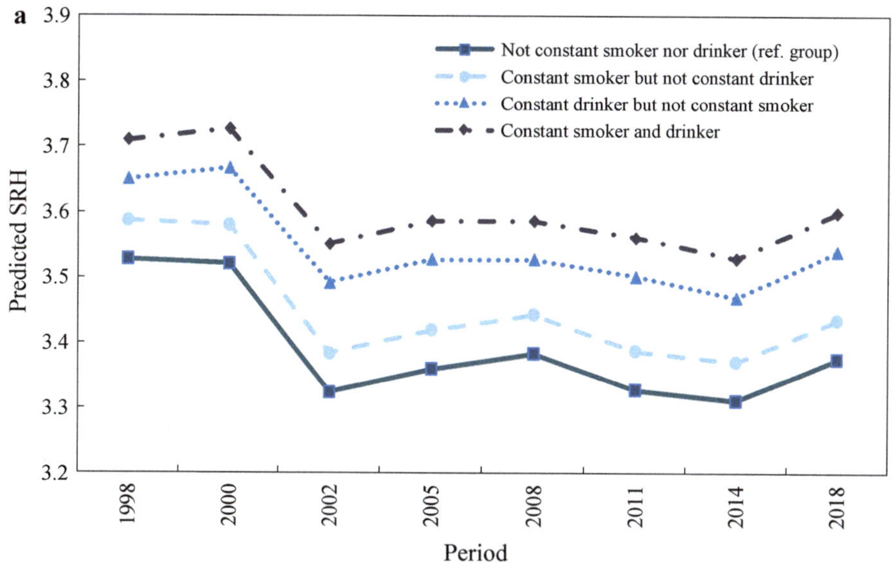

Fig. 9.2a Predicted self-rated health by smoking and drinking behaviors by time period. (Sources: 1998–2018 Chinese Longitudinal Healthy Longevity Survey (CLHLS))

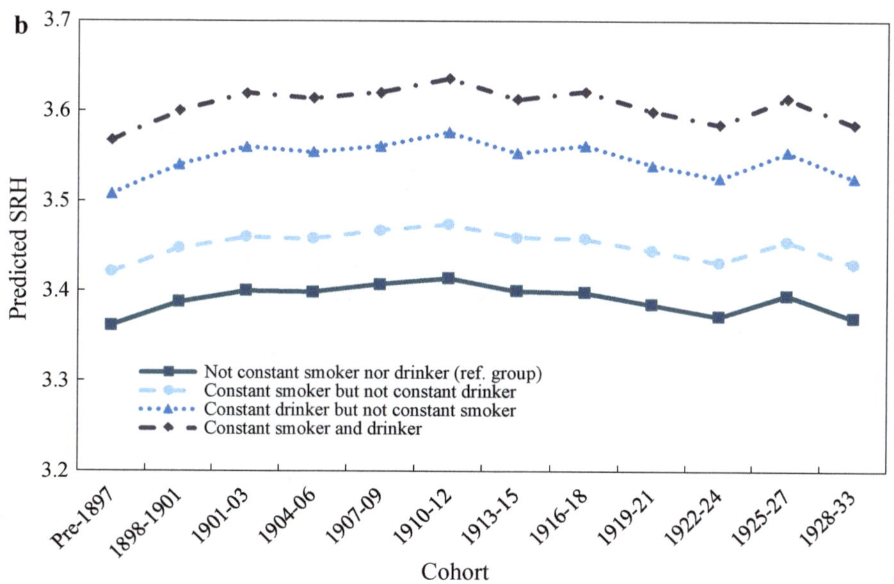

Fig. 9.2b Predicted self-rated health by smoking and drinking behaviors by birth cohort. (Sources: 1998–2018 Chinese Longitudinal Healthy Longevity Survey (CLHLS))

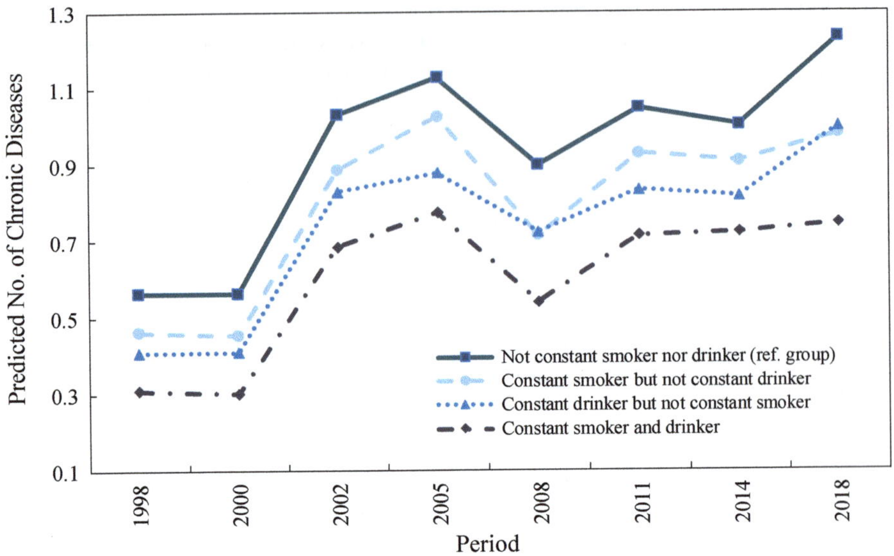

Fig. 9.3 Predicted number of chronic diseases by smoking and drinking behaviors by time period. (Sources: 1998–2018 Chinese Longitudinal Healthy Longevity Survey (CLHLS))

range from 0.5 to 1.2 for the none constant smokers or drinkers group. Over time, the predicted numbers of chronic diseases show a general increasing trend for all four groups with some crossovers being observed for a couple of groups.

Figures 9.4a and 9.4b show significant period and cohort effects when linking smoking and drinking behaviors to the oldest old's predicted cognitive function scores, controlling for demographic and socioeconomic covariates. As it is shown in Fig. 9.4a, the cognitive function score is the highest among the constant smokers and constant drinkers group, followed by the constant drinkers but not constant smokers group and the group that is composed by constant smokers but not constant drinkers. The group that is composed by not constant smokers or constant drinkers has the lowest cognitive function scores. Over time, an increasing trend is observed in predicted cognitive function scores.

When it comes to the cohort effects, interesting patterns are shown in Fig. 9.4b. As the figure shows, the health advantage in cognitive function among constant drinkers and smokers still exists and the group composed by non-constant smokers and drinkers still shows the worst cognitive function. However, crossovers occur among subgroups, indicating that the health disparities due to smoking and drinking behaviors across birth cohorts are diminishing. According to Fig. 9.4b, the 1901–03 and 1916–18 birth cohorts show the best and worst predicted cognitive function scores, respectively.

Figures 9.5 and 9.6a show significant period variations when using smoking and drinking behaviors to predict the oldest old's subjective wellbeing, measured by positive and negative feelings. Clearly, the health disparities in positive feeling

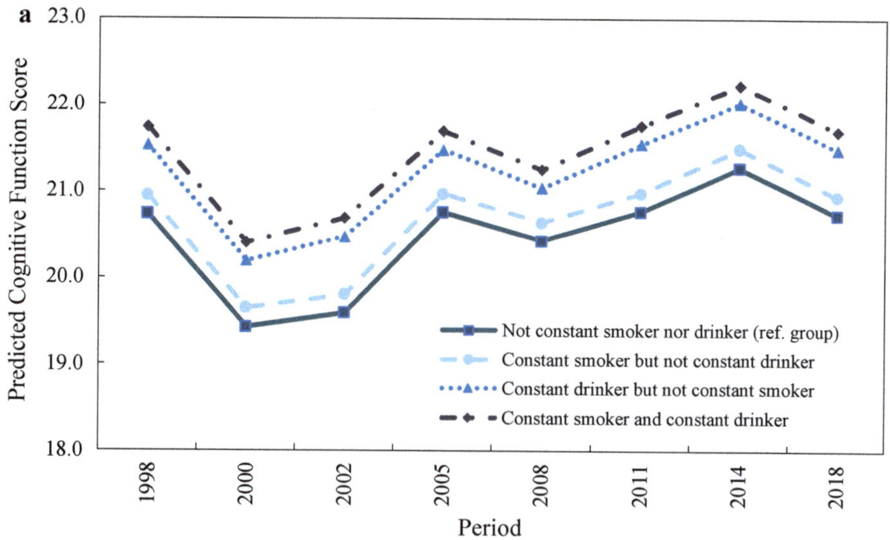

Fig. 9.4a Predicted cognitive function score by smoking and drinking behaviors by time period. (Sources: 1998–2018 Chinese Longitudinal Healthy Longevity Survey (CLHLS))

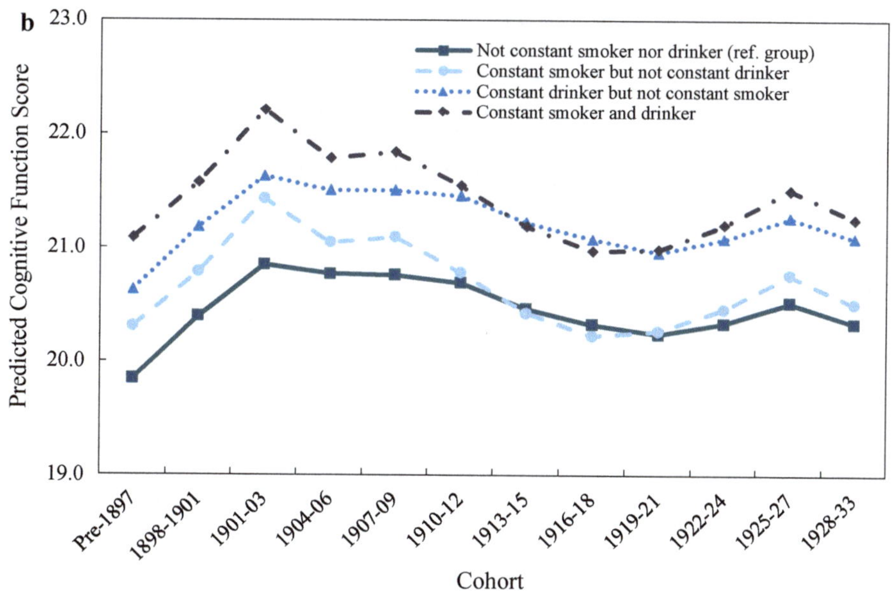

Fig. 9.4b Predicted cognitive function score by smoking and drinking behaviors by birth cohort. (Sources: 1998–2018 Chinese Longitudinal Healthy Longevity Survey (CLHLS))

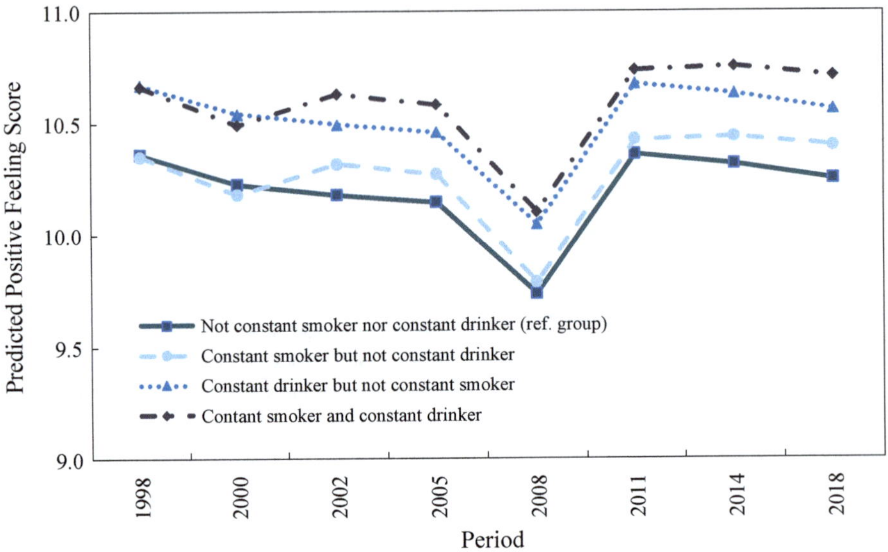

Fig. 9.5 Predicted positive feeling score by smoking and drinking behaviors by time period. (Sources: 1998–2018 Chinese Longitudinal Healthy Longevity Survey (CLHLS))

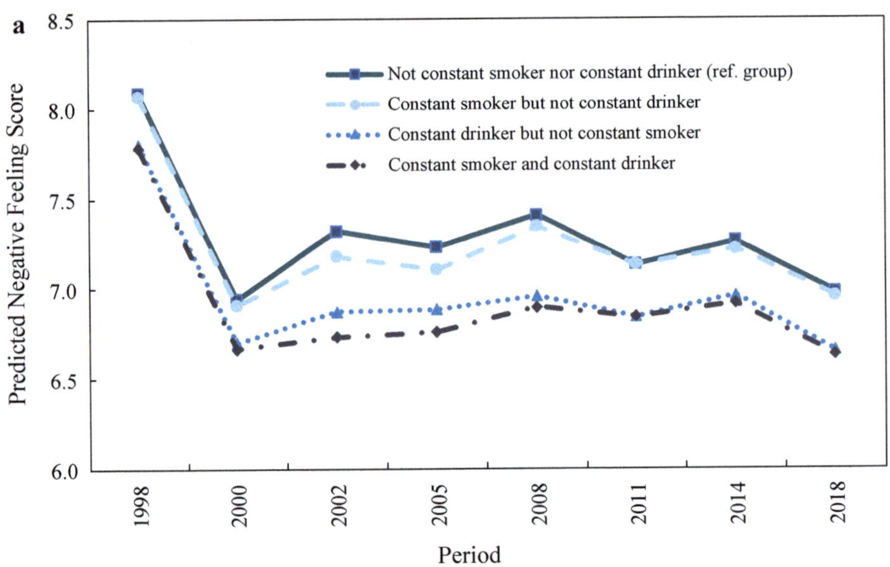

Fig. 9.6a Predicted negative feeling score by smoking and drinking behaviors by time period. (Sources: 1998–2018 Chinese Longitudinal Healthy Longevity Survey (CLHLS))

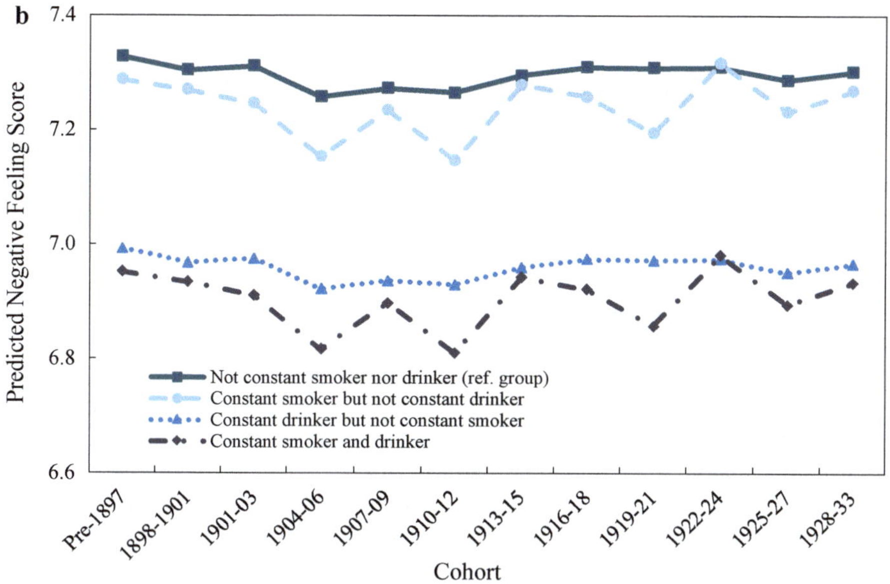

Fig. 9.6b Predicted negative feeling score by smoking and drinking behaviors by birth cohort. (Sources: 1998–2018 Chinese Longitudinal Healthy Longevity Survey (CLHLS))

scores of the four subgroups still exist. Constant smokers and drinkers demonstrate the best positive feeling scores and the lowest negative feeling scores. Similar to findings on other health measures, the health disadvantage manifested by negative feeling scores is still shown among the non-constant smokers nor drinkers group.

An apparent drop occurs in predicted positive feeling scores in year 2008, indicating significant period effects. Such a drop may well relate to the economic recession and the *Wenchuan* earthquake that occurred in that year. Regarding the predicted negative feeling scores, a general decreasing pattern is shown which suggests significant period effects.

The significant cohort effects are revealed by Fig. 9.6b. The figure shows that although constant smokers and constant drinkers still report the lowest negative feeling scores, crossovers occur between this group and the group that is composed by constant drinkers but not constant smokers. Similarly, crossovers also exist between the other two subgroups. More cohort fluctuations are observed among the group that is composed by constant smokers and constant drinkers and the group which is composed by constant smokers but not constant drinkers, indicating the subjective wellbeing of these two groups are more sensitive to cohort changes.

9.6 Conclusion and Discussion

Smoking and drinking have long been considered as important dimensions of health lifestyles that play a decisive role in shaping individual health outcomes (Daw et al., 2017; He et al., 2020; Jing et al., 2021; Lahiri & Li, 2020; Zhang et al., 2020). The existing studies on smoking and alcohol consumption behaviors and older adults' health, however, have been largely based on examining cross-sectional data. Longitudinal analyses of the compound effects of time and cohort on the oldest old's health disparities caused by their smoking and alcohol use behaviors are very limited. To fill the voids of prior literature, this chapter relies on constructing CCREMs to evaluate time-series data drawn from CLHLS on Chinese oldest old's health that span 20 years to investigate the changing patterns of Chinese oldest old's smoking and alcohol use behaviors. The study reveals the multifaceted effects of period and cohort in the associations between smoking, alcohol use and Chinese oldest old's health. Some important findings can be concluded as follows:

First, Chinese oldest old's smoking and alcohol use behaviors have been found to vary by time period and across birth cohorts. The descriptive results show that the percentage of the oldest old who smoked both in the past and at present show a declining trend over time. This is probably due to the fact that with age increasing, some elders begin to quit smoking. The descriptive results also indicate that the percentage of constant smokers has been decreasing across birth cohorts, which suggests a healthier lifestyle among more recent birth cohorts. Similar to smoking behavior, the percentage of the oldest old who constantly consumed alcohol also shows a declining trend. But no significant cohort changes are observed in such a percentage. This finding suggests that alcohol use behavior is relatively stable across birth cohorts.

The results drawn from the CCREMs also highlight the period and cohort effects when linking health disparities to the oldest old's smoking and alcohol drinking behaviors. Basically, when using smoking and alcohol drinking variables to predicted various measures of the oldest old's health, the predicted SRH and negative feeling scores of the respondents show a declining trend over time. However, the predicted number of chronic diseases and cognitive function scores demonstrate an increasing pattern by time period. The positive feeling score drops considerably in year 2008, suggesting a significant period effect caused by destructive social events, such as an economic recession and a local earthquake.

Significant cohort effects are also shown in predicted SRH, cognitive function and negative feeling scores. The results show that overall the cohort effects are not as apparent as period effects. There is no declining or increasing trend in predicted SRH or negative feeling scores. But the cognitive function scores show a general declining trend with the 1901–03 birth cohort demonstrating the best predicted cognitive function score than other birth cohorts. Crossovers are also shown in predicted cognitive function and negative feeling scores among subgroups, suggesting that the health disparities due to smoking and alcohol drinking behaviors may shrink across birth cohorts.

While demonstrating significant period and cohort effects, the study also reveals health inequalities among the oldest old with different smoking and alcohol consumption patterns. Constant smokers and drinkers show the best predicted SRH, cognitive function, subjective wellbeing and a fewer number of predicted chronic diseases. The group that smoked constantly but not drank constantly also shows health advantages, followed by the group that is composed by constant drinkers but not constant smokers. Non-constant smokers and non-constant drinkers demonstrate the worst health condition as compared to the other three subgroups. These findings seem to be contradictory to our common sense as we expect that constant smokers and drinkers to have the worst health status. The study offers several explanations for such findings.

The first possible explanation relates to the "survivor advantage" phenomenon. It is the fact that smoking and excessive alcohol drinking have detrimental effects on one's cognitive function. These behaviors also increase the risks of developing chronic diseases (Attard et al., 2017; Zhang, 2021). But since this study examines the oldest old aged 85 and over, it is reasonable to assume that those who have been heavily influenced by unhealthy lifestyles, such as drinking and smoking have passed away or have been censored. Thus, they are not included in the sample. The oldest old who were surveyed are most likely to be those that have strong survival capabilities, including the ones who claimed themselves as constant drinkers and smokers. In other words, constant drinkers and smokers may well be the ones that have the strongest survival capabilities.

The second explanation deals with the level of smoking and drinking, which has not been considered by this study. As it was stated in an earlier section of the chapter, the questions asking smoking and drinking levels in the survey contained a large number of missing values, those measures are therefore not used in this study. Thus, findings of this chapter could show biased results by only asking whether the respondents are constant smokers or drinkers. Future research may rely on measures of smoking and drinking levels to gain a more comprehensive understanding of the association between the oldest old's health and their smoking and alcohol use behaviors.

The third explanation is about the benefits of moderate alcohol drinking to individual heath. As it was pointed out in the literature review section, prior studies indicated that moderate alcohol consumption had a positive effect on health of older adults. Previous studies showed that moderate drinking has a blood pressure-lowering effect. It is also effective in reducing hypertension and anxiety, which protects cognitive function among seniors (Zhai et al., 2023). Smoking has also been found to be able to reduce pressure and anxiety, which may explain why constant smokers tend to report better SRH and subjective wellbeing. Again, including smoking and drinking levels in the analyses of older adults' health is still warranted to better understand the influence of smoking and drinking on seniors' health. Considering the above explanations, I recommend readers interpret the findings of this chapter with caution.

To conclude, this chapter has investigated the period and cohort effects in the relationship between Chinese oldest old's smoking and alcohol drinking behaviors and their health status. The research also demonstrates health disparities due to smoking and alcohol drinking behaviors. When using smoking and alcohol use behaviors to predict Chinese oldest old's health status, significant period and cohort effects are shown with period effects being more prominent than cohort effects. These findings have enriched the existing analyses and theories on health lifestyles and the oldest old's health. Future research may consider incorporating smoking and drinking levels to expand the current study. Scholars may also explore social, economic, political and cultural factors that may have caused the period and cohort variations when linking smoking and alcohol consumption to the oldest old's health.

References

Agnihotri, R., & Gaur, S. (2014). Implications of tobacco smoking on the oral health of older adults. *Geriatrics & Gerontology International, 14*(3), 526–540.

Almeida De, N. M. G., Nascimento-Souza, M. A., Lima-Costa, M. F., & Peixoto, S. V. (2020). Lifestyle factors and multimorbidity among older adults (ELSI-Brazil). *European Journal of Ageing, 17*, 521–529.

Ambati, S. K. (2021). Association of passive smoking and lung cancer: Is there substantial evidence to prove the causation? *International Journal of Community Medicine and Public Health, 8*(5), 2540–2543.

Araújo, J., Ramos, E., & Lopes, C. (2011). Lifestyles and self-rated health, in Portuguese elderly from rural and urban area (in Portuguese). *Acta Médica Portuguesa, 24*, 79–88. https://doi.org/10.20344/amp.1524

Aspinall, R., & Lang, P. O. (2018). Interventions to restore appropriate immune function in the elderly. *Immunity & Ageing, 15*(1), 1–8.

Attard, R., Dingli, P., Doggen, C. J., Cassar, K., Farrugia, R., & Wettinger, S. B. (2017). The impact of passive and active smoking on inflammation, lipid profile and the risk of myocardial infarction. *Open Heart, 4*(2), e000620.

Barengo, N. C., Antikainen, R., Harald, K., & Jousilahti, P. (2019). Smoking and cancer, cardiovascular and total mortality among older adults: The Finrisk study. *Preventive Medicine Reports, 14*, 100875.

Behl, T. A., Stamford, B. A., & Moffatt, R. J. (2023). The effects of smoking on the diagnostic characteristics of metabolic syndrome: A review. *American Journal of Lifestyle Medicine, 17*(3), 397–412.

Burns, D. M. (2000). Cigarette smoking among the elderly: Disease consequences and the benefits of cessation. *American Journal of Health Promotion, 14*(6), 357–361.

Cao, B. (2016). Future healthy life expectancy among older adults in the US: A forecast based on cohort smoking and obesity history. *Population Health Metrics, 14*(1), 1–14.

Chen, S. F., Wu, P., Zhou, L., Shen, Y. Q., Li, Y. J., & Song, H. M. (2015). Relationship between increase of serum homocysteine caused by smoking and oxidative damage in elderly patients with cardiovascular disease. *International Journal of Clinical and Experimental Medicine, 8*(3), 4446–4454.

Chen, H., Zhang, X., Feng, Q., & Zeng, Y. (2022). The effects of "diet–smoking–gender" three-way interactions on cognitive impairment among Chinese older adults. *Nutrients, 14*(10), 2144.

Choi, N. G., & DiNitto, D. M. (2021). Alcohol use disorder and treatment receipt among individuals aged 50 years and older: Other substance use and psychiatric correlates. *Journal of Substance Abuse Treatment, 131*, 108445.

Daw, J., Margolis, R., & Wright, L. (2017). Emerging adulthood, emerging health lifestyles: Sociodemographic determinants of trajectories of smoking, binge drinking, obesity, and sedentary behavior. *Journal of Health and Social Behavior, 58*, 181–197.

Domingo-Relloso, A., Joehanes, R., Rodriguez-Hernandez, Z., Lahousse, L., Haack, K., & Fallin, M. D. (2023). Smoking, blood DNA methylation sites and lung cancer risk. *Environmental Pollution, 334*, 122153.

Escobedo, L. G., Reddy, M., & DuRant, R. H. (1997). Relationship between cigarette smoking and health risk and problem behaviors among US adolescents. *Archives of Pediatrics & Adolescent Medicine, 151*(1), 66–71.

Gross, A. L., Rebok, G. W., Ford, D. E., Chu, A. Y., Gallo, J. J., & Liang, K. Y. (2011). Alcohol consumption and domain-specific cognitive function in older adults: Longitudinal data from the Johns Hopkins precursors study. *Journals of Gerontology Series B: Psychological Sciences and Social Sciences, 66*(1), 39–47.

Han, B. H., Moore, A. A., Sherman, S. E., & Palamar, J. J. (2018). Prevalence and correlates of binge drinking among older adults with multimorbidity. *Drug and Alcohol Dependence, 187*, 48–54.

Harris, K. K., Zopey, M., & Friedman, T. C. (2016). Metabolic effects of smoking cessation. *Nature Reviews Endocrinology, 12*(5), 299–308.

He, F., Li, T., Lin, J., Li, F., Zhai, Y., & Zhang, T. (2020). Passive smoking exposure in living environments reduces cognitive function: A prospective cohort study in older adults. *International Journal of Environmental Research and Public Health, 17*(4), 1402.

Hecht, S. S. (2006). Smoking and lung cancer--a new role for an old toxicant? *Proceedings of the National Academy of Sciences of the United States of America, 103*(43), 15725–15726.

Hou, W., Chen, S., Zhu, C., Gu, Y., Zhu, L., & Zhou, Z. (2023). Associations between smoke exposure and osteoporosis or osteopenia in a US NHANES population of elderly individuals. *Frontiers in Endocrinology, 14*, 1074574.

Huang, W. C., Huang, Y. C., Lee, M. S., Chang, H. Y., & Doong, J. Y. (2021). Frailty severity and cognitive impairment associated with dietary diversity in older adults in Taiwan. *Nutrients, 13*(2), 418.

Jefferis, B. J., Lawlor, D. A., Ebrahim, S., Wannamethee, S. G., Feyerabend, C., Doig, M., & Whincup, P. H. (2010). Cotinine-assessed second-hand smoke exposure and risk of cardiovascular disease in older adults. *Heart, 96*(11), 854–859.

Jeong, K., Cho, S., Ryu, J., Cho, H. J., & Kim, S. (2022). Effects of changes in smoking behavior of older adults'oral health. *Healthcare, 10*(11), 2127.

Jing, Z., Li, J., Wang, Y., Yuan, Y., Zhao, D., & Hao, W. (2021). Association of smoking status and health-related quality of life: Difference among young, middle-aged, and older adults in Shandong, China. *Quality of Life Research, 30*, 521–530.

Kihara, T., Yamagishi, K., Iso, H., Tamakoshi, A., & JACC Study Group. (2022). The association of smoking cessation with mortality from pneumonia among middle-aged and elderly community residents: The Japan Collaborative Cohort (JACC) study. *Preventive Medicine, 163*, 107188.

Kunzmann, A. T., Coleman, H. G., Huang, W. Y., & Berndt, S. I. (2018). The association of lifetime alcohol use with mortality and cancer risk in older adults: A cohort study. *PLoS Medicine, 15*(6), e1002585.

Lahiri, K., & Li, X. (2020). Smoking behavior of older adults: A panel data analysis using HRS. *Journal of Quantitative Economics, 18*, 495–523.

Lam, T. H., Xu, L., Schooling, C. M., Chan, W. M., Lee, S. Y., & Leung, G. M. (2015). Smoking and mortality in a prospective cohort study of elderly C hinese in Hong Kong. *Addiction, 110*(3), 502–510.

Lau, E. M. C., Suriwongpaisal, P., Lee, J. K., Das De, S., Festin, M. R., & Saw, S. M. (2001). Risk factors for hip fracture in Asian men and women: The Asian osteoporosis study. *Journal of Bone and Mineral Research, 16*(3), 572–580.

Lesnoff-Caravaglia, G. (2007). *Health aspects of aging: The experience of growing old.* Charles C Thomas Publisher.

Lewis, T. (2008). Assessing an older adult for alcohol use disorders. *Nursing 2023, 38*(6), 60–61.
Li, K., Yao, C., Di, X., Yang, X., Dong, L., Xu, L., & Zheng, M. (2016). Smoking and risk of all-cause deaths in younger and older adults: A population-based prospective cohort study among Beijing adults in China. *Medicine, 95*(3).
Lim, S. J., Gombojav, B., Jee, S. H., Nam, C. M., & Ohrr, H. (2012). Gender-specific combined effects of smoking and hypertension on cardiovascular disease mortality in elderly Koreans: The Kangwha cohort study. *Maturitas, 73*(4), 331–336.
Lim, S. H., Tai, B. C., Yuan, J. M., Mimi, C. Y., & Koh, W. P. (2013). Smoking cessation and mortality among middle-aged and elderly Chinese in Singapore: The Singapore Chinese health study. *Tobacco Control, 22*(4), 235–240.
Lu, Y., Sugawara, Y., Zhang, S., Tomata, Y., & Tsuji, I. (2020). Smoking cessation and incident dementia in elderly Japanese: The Ohsaki cohort 2006 study. *European Journal of Epidemiology, 35*, 851–860.
Mons, U., Schöttker, B., Müller, H., Kliegel, M., & Brenner, H. (2013). History of lifetime smoking, smoking cessation and cognitive function in the elderly population. *European Journal of Epidemiology, 28*, 823–831.
Mukamal, K. J., Robbins, J. A., Cauley, J. A., Kern, L. M., & Siscovick, D. S. (2007). Alcohol consumption, bone density, and hip fracture among older adults: The cardiovascular health study. *Osteoporosis International, 18*, 593–602.
Nash, S. H., Liao, L. M., Harris, T. B., & Freedman, N. D. (2017). Cigarette smoking and mortality in adults aged 70 years and older: Results from the NIH-AARP cohort. *American Journal of Preventive Medicine, 52*(3), 276–283.
Ng, R., Sutradhar, R., Yao, Z., Wodchis, W. P., & Rosella, L. C. (2020). Smoking, drinking, diet and physical activity—Modifiable lifestyle risk factors and their associations with age to first chronic disease. *International Journal of Epidemiology, 49*(1), 113–130.
Ortolá, R., García-Esquinas, E., Galán, I., Guallar-Castillón, P., López-García, E., Banegas, J. R., & Rodríguez-Artalejo, F. (2017). Patterns of alcohol consumption and risk of falls in older adults: A prospective cohort study. *Osteoporosis International, 28*, 3143–3152.
Ostbye, T., & Taylor, D. H. (2004). The effect of smoking on years of healthy life (YHL) lost among middle-aged and older Americans. *The effect of smoking on years of healthy life (YHL) lost among middle-aged and older Americans, 39*(3), 531–552.
Otsuka, R., Nishita, Y., Tange, C., Tomida, M., Kato, Y., Nakamoto, M., & Shimokata, H. (2017). Dietary diversity decreases the risk of cognitive decline among Japanese older adults. *Geriatrics & Gerontology International, 17*(6), 937–944.
Pan, X. F., Li, Y., & Franco, O. H. (2020). Impact of combined lifestyle factors on all-cause and cause-specific mortality and life expectancy in Chinese: The Singapore Chinese health study. *The Journals of Gerontology. Series A, Biological Sciences and Medical Sciences, 75*, 2193–2199. https://doi.org/10.1093/gerona/glz271
Panza, F., Capurso, C., D'Introno, A., Colacicco, A. M., Frisardi, V., & Lorusso, M. (2009). Alcohol drinking, cognitive functions in older age, predementia, and dementia syndromes. *Journal of Alzheimer' s Disease, 17*(1), 7–31.
Peltzer, K., & Phaswana-Mafuya, N. (2013). Problem drinking and associated factors in older adults in South Africa. *African Journal of Psychiatry, 16*(2), 104–109.
Piano, M. R., Mazzuco, A., Kang, M., & Phillips, S. A. (2017). Cardiovascular consequences of binge drinking: An integrative review with implications for advocacy, policy, and research. *Alcoholism: Clinical and Experimental Research, 41*(3), 487–496.
Sacco, P., Unick, G. J., Kuerbis, A., Koru, A. G., & Moore, A. A. (2015). Alcohol-related diagnoses in hospital admissions for all causes among middle-aged and older adults: Trends and cohort differences from 1993 to 2010. *Journal of Aging and Health, 27*(8), 1358–1374.
Sanada, F., Taniyama, Y., Muratsu, J., Otsu, R., Shimizu, H., Rakugi, H., & Morishita, R. (2018). Source of chronic inflammation in aging. *Frontiers in Cardiovascular Medicine, 5*, 12.
Sarkar, D., Jung, M. K., & Wang, H. J. (2015). Alcohol and the immune system. *Alcohol Research: Current Reviews, 37*(2), 153.

Sarrio, R. G., Torregrosa, M. D., Lopez, P., Gomez-Codina, J., & Rosell, R. (2010). Smoking habits in elderly lung cancer patients: Still no changes in epidemiology? A single-center experience. *Clinical & Translational Oncology, 12*(10), 686–691.

Saunders, P. A., Copeland, J. R., Dewey, M. E., Davidson, I. A., Mcwilliam, C., Sharma, V., & Sullivan, C. (1991). Heavy drinking as a risk factor for depression and dementia in elderly men: Findings from the Liverpool longitudinal community study. *The British Journal of Psychiatry, 159*(2), 213–216.

Shimizu, Y., Nakazato, M., Sekita, T., Kadota, K., Sato, S., & Koyamatsu. (2014). Associations between body height and drinking status with diabetes in a rural, non-overweight, elderly, Japanese male population: The Nagasaki Islands study. *Journal of Diabetes Metabolism, 5*(339), 2.

Strine, T. W., Okoro, C. A., Chapman, D. P., Balluz, L. S., Ford, E. S., & Ajani, U. A. (2005). Health-related quality of life and health risk behaviors among smokers. *American Journal of Preventive Medicine, 28*(2), 182–187.

Tamaki, J., Iki, M., Fujita, Y., Kouda, K., Yura, A., & Kadowaki. (2011). Impact of smoking on bone mineral density and bone metabolism in elderly men: The Fujiwara-kyo Osteoporosis Risk in Men (FORMEN) study. *Osteoporosis International, 22*, 133–141.

Taylor, D. H., Jr., Hasselblad, V., Henley, S. J., Thun, M. J., & Sloan, F. A. (2002). Benefits of smoking cessation for longevity. *American Journal of Public Health, 92*(6), 990–996.

Tolstrup, J. S., Hvidtfeldt, U. A., Flachs, E. M., Heitmann, B. L., Bälter, K., & Goldbourt, U. (2010). Smoking and risk of coronary heart disease in younger, middle-aged, and older adults. *Longitudinal and Life Course Studies, 1*(3), 131.

Tolstrup, J. S., Hvidtfeldt, U. A., Flachs, E. M., Spiegelman, D., Heitmann, B. L., Bälter, K., & Feskanich, D. (2014). Smoking and risk of coronary heart disease in younger, middle-aged, and older adults. *American Journal of Public Health, 104*(1), 96–102.

Traber, M. G., van der Vliet, A., Reznick, A. Z., & Cross, C. E. (2000). Tobacco-related diseases: Is there a role for antioxidant micronutrient supplementation? *Clinics in Chest Medicine, 21*(1), 173–187.

Wang, M., Luo, X., Xu, S., Liu, W., Ding, F., Zhang, X., & Wang, W. (2019). Trends in smoking prevalence and implication for chronic diseases in China: Serial national cross-sectional surveys from 2003 to 2013. *The Lancet Respiratory Medicine, 7*(1), 35–45.

World Health Organization. (2021). *WHO report on the global tobacco epidemic 2021: Addressing new and emerging products*. Retrieved from https://www.who.int/en/news-room/fact-sheets/detail/tobacco.

Yang, Y., & Land, K. C. (2006). A mixed models approach to the age-period-cohort analysis of repeated Cross-section surveys, with an application to data on trends in verbal test scores. *Sociological Methodology, 36*(1), 75–97.

Yang, L., & Zikos, V. (2023). Mental health and smoking behavior. *Economic Modelling, 106407*.

Zammit, A. R., Katz, M. J., Derby, C., Bitzer, M., & Lipton, R. B. (2016). Metabolic syndrome and smoking are associated with future development of advanced chronic kidney disease in older adults. *Cardiorenal medicine, 6*(2), 108–115.

Zeng, Y., Vaupel, J., Xiao, Z., Zhang, C., & Liu, Y. (2002). Sociodemographic and health profiles of the oldest old in China. *Population and Development Review, 28*, 251–273.

Zhai, Y., Cui, C., Yu, T., & Su, Z. (2023). Epidemiological survey of hypertension among middle-aged and elderly adults in Anhui Province. *Medicine, 102*(30), e34402.

Zhang, L. (2021). A latent class analysis of health lifestyles and health outcomes among Chinese older adults. *Ageing & Society, 41*, 313–338.

Zhang, R., Shen, L., Miles, T., Shen, Y., & Li, C. (2020). Association of low to moderate alcohol drinking with cognitive functions from middle to older age among us adults. *JAMA Network Open, 3*(6).

Conclusion

10

Global aging has made health related issues become a major concern in world countries since the majority of the challenges aging brings to people and the society largely link to health issues. Health condition greatly impacts older adults' wellbeing and their quality of life. Since the oldest old is a fast growing population group, health issues of the oldest old are more prominent than ever. Thus, it is essential to study influential factors of the oldest old's health, investigate solutions for treatment, and provide health care and effective disease management services to the oldest old. Such practices can reduce the need for hospitalization and emergency visits, which alleviates the burden on healthcare systems, the society as well as families.

China, a country that has practiced the one-child policy for 34 years, is now experiencing serious challenges of population aging due to low fertility. According to the Chinese culture, longevity, happiness, wealth, good fortune, and prosperity are considered as five fortunes for thousands of years. Longevity has long been a pursuit of the Chinese population. Under such a cultural background, maintaining healthy condition and pursuing longevity are extremely important to Chinese. Similar to other developed countries, the oldest old population is growing rapidly in China. This group has a high demand on long-term care and medical resources. To meet the demands of this group, in addition to establishing nation-wide health care systems, it is also important to explore the changing patterns of the oldest old's health and its determinants and launch effective disease prevention strategies to promote the oldest old's health. This becomes urgent tasks in the years to come.

In the academic world, since the late twentieth century, a considerable amount of attention has been paid to the study of older adults' health from multiple disciplines. Abundant literature has been documented in the fields of biology, psychology, medicine, epidemiology and so forth. Sociologists and demographers, in particular, have concentrated on evaluating older adults' (including the oldest old's) health and its relationships with a variety of social, behavioral and demographic factors, such as social environment, dietary patterns, social participation, internet use, grandparental

L. Zhang, *Living Longer and Healthier at Older Ages*, Advances in Studies of Aging and Health 3, https://doi.org/10.1007/978-3-031-69773-9_10

care, intergenerational support, living arrangement and health lifestyles (Chai & Mei, 2022; Chen & Silverstein, 2000; Lee et al., 2013; Luo et al., 2020; Phulkerd et al., 2020; Rizzuto et al., 2019; Saha et al., 2022; Tang et al., 2020; Wang, 2022; Wang et al., 2020; Wei et al., 2022; Xu & Zhang, 2023; Yang et al., 2021; Yang et al., 2022; Yuk et al., 2021; H. Zhang et al., 2022b; Zhang et al., 2021; Zhang, 2021; Zhao et al., 2021a; Zhao et al., 2021b; Zhou et al., 2017). These studies, however, are mostly cross-sectional in nature. Only very limited research has applied a longitudinal scope to shed light on the changing patterns of the oldest old's health and its determinants. By reviewing prior literature, Chap. 2 has pointed out that as to the case of China, the existing longitudinal studies have mainly examined Chinese older adults' trajectories and trends of frailty, chronic disease incidence rates, cognitive function change and mortality rates (Aihemaitijiang et al., 2022; Guo et al., 2022; Hou et al., 2018; Ren et al., 2022; Yu et al., 2018).

Even with these analyses on health of older adults (including the oldest old), few simultaneous assessments of the age, time period, and birth cohort effects are found in the analyses of health disparities among Chinese oldest old. Thus far, we still know very little about the compound effects of age, period and cohort on Chinese oldest old's health. Therefore, the aim of this book is to use the age-period-cohort (APC) approach to examine how Chinese oldest old's health has changed over time and across birth cohorts. The research also aims to elucidate how health trajectories of the Chinese oldest old are shaped by a variety of factors across different birth cohorts, and how historical events may leave imprints on cohorts' life experiences so that diverse health outcomes have occurred. This is one of the first studies that consider the ever changing social contexts and bring a longitudinal scope to the research on Chinese oldest old's health. This exercise improves our understanding on Chinese oldest old's health and its determinants. Findings of this manuscript also provide evidence to enrich gerontology theories on health.

The manuscript begins by providing a synthetic review of prior literature on health and its determinants among older adults, particularly the oldest old. The topics reviewed are from multiple disciplines. The chapter then pays a particular attention to sociological and demographic studies of health among older adults, including the oldest old. The reviewed health determinants include demographic and socioeconomic factors, intergenerational support, government and community support, healthy lifestyles, physical exercise, living arrangements, grandparental child care, internet use et al. I have emphasized that previous studies have significantly advanced our understanding on health of older adults (including the oldest old) and its influential factors. Nonetheless, most existing studies have applied a cross-sectional scope, which prevents us from understanding the dynamic patterns of the oldest old's health and its determinants. The age, period and cohort effects are still unclear when linking some important demographic, behavioral and social factors to the oldest old's health, which remains a gap in the existing literature. This manuscript thus intends to improve previous literature by filling such voids.

After reviewing prior literature in Chaps. 2 and 3 introduces data and methods used in this manuscript. The chapter has introduced the concepts of period and birth cohort as well as the period and cohort effects. The chapter also points out the strengths of applying the APC approach when studying the oldest old's health. Basically, the APC approach has solved the underidentification problem that existed in previous studies (Yang & Land, 2006). The CCREM used in this study takes the multilevel structure of the data design into consideration. That way, the compound effects of age, period and cohort can be detected. The chapter also introduces several datasets that allow researchers to conduct longitudinal analyses of Chinese older adults' health, including the Chinese Longitudinal Healthy Longevity Survey (CLHLS) data that are analyzed in this manuscript.

Chapters 4, 5, 6, 7, 8 and 9 start to examine the changing patterns of Chinese oldest old's health and its determinants by analyzing data from eight (1998, 2000, 2002, 2005, 2008, 2011–12, 2014 and 2018) waves of the Chinese Longitudinal Healthy Longevity Survey (CLHLS). Chapter 4 describes how Chinese oldest old's health, measured by SRH, chronic diseases, cognitive function and subjective well-being, has changed by time period and across birth cohorts. Chapters 5 and 6 examine demographic and socioeconomic factors and their associations with the oldest old's health. Chapters 7 and 8 shift the research interests to physical exercise and leisure activities and their influence on the oldest old's health outcomes. Chapter 9 investigates the smoking and alcohol use behaviors of Chinese oldest old and how period and cohort effects play a role in the linkages between smoking, alcohol use and the oldest old's health. By examining the Chinese sample, the monograph has revealed how a variety of demographic, socioeconomic, behavioral and lifestyle factors have affected the oldest old's health status and has elucidated significant period and cohort effects. Chapter 10 concludes findings and tries to bring together several areas of the book to illuminate how this manuscript improves our understanding of the oldest old's health from a longitudinal point of view. The ending remarks further place future research attention on applying a longitudinal scope to scientific studies of the oldest old's health related issues. In the following subsections, I highlight key findings of the book.

10.1 Summary of Findings

10.1.1 Changing Patterns of Chinese Oldest Old's Health Outcomes by Period and Birth Cohort

10.1.1.1 Statistical Results

The manuscript first presents the changing patterns of Chinese oldest old's health by time period and across birth cohorts through analyzing descriptive results, which is shown in Chap. 4. Findings reveal that regarding self-rated health (SRH), the majority of the respondents rated their overall health as "fair" (35.7%) or "good" (38.7%). Less than one-fifth of the respondents rated their health as "very good". Except for the 1928 to 1938 birth cohort, younger birth cohorts seem to be more inclined to self-rate their health as "very good" as compared to older birth cohorts. As to the period variations, although some fluctuations occur, a general declining trend is shown during 1998 to 2014. That is, before year 2014, for a specific birth cohort, with time goes, the percentage of the oldest old who reported their health as "very good" tends to decline. Since 2014, the percentages of respondents who rated their health as "very good" begins to increase for several birth cohorts.

For the second health measure, average number of chronic diseases, the values range from 0.6 in 1998 to 1.6 in 2018 with an increasing trend over time and across birth cohorts. The differentials become more significant since year 2008 with more recent birth cohorts reporting greater average numbers of chronic diseases as compared to earlier birth cohorts. As to variations in chronic diseases by time period, it appears that respondents in years 2002 to 2005 reported greater average numbers of chronic diseases for various birth cohorts. The average numbers of chronic diseases then decline in years 2006 to 2008; the numbers rise again since year 2008.

Cognitive function is the third health measure and its average values range between 16.1 to 26.1. Younger birth cohorts show a better cognitive function status than older birth cohorts. With time goes, the oldest old's cognitive function scores show a declining pattern. It makes sense because the studied respondents are aging over time and their cognitive function deteriorates by time period.

The average positive feeling scores for the respondents of various birth cohorts range from 9.5 to 11.2, indicating a fairly good subjective wellbeing status of the studied oldest old. No significant birth cohort differentials are shown; and more recent birth cohorts only report slightly higher average positive feeling scores than earlier birth cohorts. As far as variations in time periods, it shows that before year 2008, the average positive feeling scores have been declining over time. After 2008, the mean positive feeling scores for various birth cohorts increase and reach their peaks in year 2011 for most birth cohorts. The lowest positive feeling scores shown in year 2008 may well relate to the *Wenchuan* earthquake and the economic depression that China experienced during that time.

The average negative feeling scores are from 6.5 to 8.0 for the studied birth cohorts. The average negative feeling scores reported by the respondents suggest a fairly good subjective wellbeing status of the studied Chinese oldest old. No significant cohort differences are observed in this case. When the time effect is considered, years 2000 and 2018 are the periods that show the lowest average negative feeling scores. Years 2002, 2008 and 2014 appear to show higher negative feeling scores for the sampled birth cohorts. In short, period variations are more apparent than cohort variations in terms of Chinese oldest old's negative feeling scores. These findings suggest that Chinese oldest old's subjective wellbeing is more likely to be affected by period effects due to social changes than by birth cohort variations.

Overall, the findings show that Chinese oldest old's SRH, negative feeling scores and cognitive function scores tend to decline over time. But the average number of chronic diseases increases over time. The positive feeling scores show a "V" shape, indicating significant period influence on Chinese oldest old's subjective wellbeing. Cohort wise, the average number of chronic diseases and cognitive function scores are in a rising pattern across birth cohorts. The cohort variations are not as apparent as the period variations for other health measures.

10.1.1.2 Basic CCREM Results

The manuscript then applies the APC approach through constructing CCREMs by only including period, cohort, age and age-squared term. The results from CCREMs indicate the overall impacts of age, period and cohort on health measures. The age effect appears to be significant but the quadratic age effect on the oldest old's health seems to be non-significant net of the random period and cohort effects. It implies that the relationships between age and health measures are linear. That is, adjusting for time period and birth cohort variations, with age increasing, the level of SRH increases but the levels of other health measures decline.

The estimates of random effects in terms of residual variance components at level-2 suggest significant period and cohort effects when controlling for the age effect. The results show that the predicted SRH varies significantly by time period and across birth cohorts. The period effects exhibit clear nonlinear declines over time. Years 2002 and 2014 appear to be the years that show lower predicted SRH among all time periods. The cohort effects first show an increasing trend prior to the 1910 to 1912 cohort, and a declining trend is observed for later birth cohorts. When predicting the number of chronic diseases, significant period and cohort effects are detected. The results show that the predicted number of chronic diseases changes from 0.6 in 1998 to 1.4 in 2018, indicating an increasing trend over time. The cohort effects are mainly demonstrated by a downward trend for earlier birth cohorts in terms of chronic diseases. An upward trend is then observed for the 1916–18 and later birth cohorts. Interestingly, a curvy pattern instead of a rising trend in chronic diseases is observed according to the results from CCREMs.

As to cognitive function, significant period and cohort effects are shown as well. The period effects display a downward trend during 1998 to 2000. Since year 2000, a general upward trend is observed over time although fluctuations are shown in some years. Overall, the results suggest that the predicted cognitive function of the oldest old is getting better over time and younger birth cohorts have better cognitive function than older birth cohorts.

Significant period and cohort effects are also observed when predicting positive feeling scores. The most intriguing finding is the dip in predicted positive feeling scores for year 2008. That was the year when world countries, including China, experienced an economic hardship. These findings indicate that the predicted positive feeling scores are fairly stable over time except for the year that China experienced an economic recession and the *Wenchuan* earthquake. The 1898–1900 and 1925–27 birth cohorts seem to be the ones that report better positive feelings than other birth cohorts. The predicted positive feeling scores tend to decline across birth cohorts, suggesting younger birth cohorts are less positive than older birth cohorts.

A declining trend is also observed in terms of predicted negative feeling scores, implying that Chinese oldest old's subjective wellbeing that is measured by negative feeling scores is getting better over time. The predicted negative feeling scores display a curvy trend across birth cohorts with the 1904–06 birth cohort showing the lowest predicted negative feeling scores than other birth cohorts.

10.1.2 Significant Period and Cohort Effects in the Associations of the Oldest Old's Health and a Variety of Factors

Chapters 5, 6, 7, 8 and 9 present the CCREM results based on examining the associations between Chinese oldest old's health and a variety of demographic, socioeconomic, behavioral and lifestyle factors. Significant period and cohort effects are demonstrated in each chapter. Chapter 5 finds that demographic characteristics have significant impacts on the oldest old's health. Demographic factors work in the way as demographic and sociological theories expect. Men on average report a lower number of chronic diseases and show better SRH, cognitive function and subjective wellbeing as compared to women. Married individuals show better cognitive function and subjective wellbeing, but they report worse SRH and a greater number of chronic diseases than those who were divorced, widowed or separated.

The study then puts each of the demographic variables into the second level of the model to investigate the period and cohort effects when controlling for the SES covariates. The most important findings are about the period and cohort variations in the linkage between demographic factors and the oldest old's health. The results show that when *sex* is used to predict the oldest old's health, significant period and cohort effects are shown when estimating SRH, chronic diseases and cognitive function. Significant period effects are found when predicting positive feeling scores and significant cohort effects are observed when estimating negative feeling scores. In terms of the linkage between *residence* and health, both period and cohort effects are found when predicting SRH, chronic diseases and cognitive function. Significant period effects are found when evaluating the subjective wellbeing measures. As to *marital status*, significant period and cohort effects are observed when predicting chronic diseases, cognitive function and negative feeling scores. Only period effects are observed when predicting SRH and positive feeling scores.

When specific period and cohort effects are taken into consideration, the findings suggest that health outcomes of males, rural residents and unmarried respondents are more sensitive to cohort and period variations than their female, urban and married counterparts. The most significant period effect is shown when predicting positive feeling scores of the oldest old. There is a sharp decline in year 2008 when using each demographic variable to estimate the respondent's positive feeling scores. Again, this is likely to be caused by the economic hardship and the *Wenchuan* earthquake that occurred in that particular year. The chapter concludes that unlike physical health, the oldest old's subjective wellbeing is more likely to be influenced by certain social events occurred in unique periods.

Chapter 6 studies socioeconomic factors and the oldest old's health. The results fall in line with findings from cross-sectional analyses that socioeconomic factors are major determinants of the oldest old's health outcomes, controlling for demographic characteristics. The oldest old who have higher educational attainments and seldom went to bed hungry in childhood show better health outcomes except for chronic diseases. These findings suggest that higher SES is positively associated with the oldest old's health status. Individual SES status in childhood has a cumulative effect on his or her health in older ages, which supports the cumulative advantage/disadvantage theory.

The most essential findings in Chap. 6 are the period and cohort variations in the linkage between SES and the oldest old's health. In general, the findings suggest that health of individuals with a lower SES is more sensitive to cohort and period changes as compared to those with a higher SES. The most significant period effect is shown when predicting positive feeling scores of the oldest old. There is a dip when using each SES variable to estimate the respondent's positive feeling scores. Again, this relates to the economic hardship and the *Wenchuan* earthquake that occurred in that particular year. Findings from this chapter once again highlight that social events have stronger influence on the oldest old's subjective wellbeing than on their physical health.

Chapter 7 focuses on examining the period and cohort variations in Chinese oldest old's physical activity patterns and their health outcomes during the studied 20-year span. The percentage of the oldest old who exercised both in the past and at present show a declining trend over time due to aging. But such a percentage seems to increase among younger birth cohorts, indicating that younger birth cohorts are more active than older ones. The health disparities predicted by the oldest old's physical activity variable also vary by period and cohort. When using the physical activity variable to predict the oldest old's health, those who exercised both in the past and at present have better SRH, cognitive function, subjective wellbeing and a fewer number of chronic diseases than the reference group that is more sedentary. The predicted cognitive function tends to increase over time. The positive feeling scores show a significant drop in year 2008; and the predicted SRH shows a declining trend over time. Significant cohort changes are observed when predicting SRH, cognitive function, chronic diseases and negative feeling scores. In addition, the cognitive function advantage among the more active subgroup is declining across birth cohorts.

Through evaluating the link between leisure activities and the oldest old's health, findings of Chap. 8 show significant period and cohort variations in Chinese oldest old's leisure activity patterns during the studied periods. The descriptive results indicate that the percentage of the oldest old who engaged in at least one leisure activity daily has been declining over time. This this because the respondents were getting older over time and they became less active in participating in leisure activities. But the cohort differentials are more apparent when it comes to the leisure activity measure. Such findings imply that younger birth cohorts become more active, especially in leisure activities.

Moreover, the health disparities predicted by the oldest old's leisure activity patterns vary by period and cohort. The results suggest that those who engaged in at least one leisure activity almost every day show better SRH, cognitive function and subjective wellbeing. But they tend to have a greater number of chronic diseases than those who did not engage in any leisure activities on a daily basis. Over time, when using the leisure activity variable to predict the oldest old's health status, the predicted SRH shows a declining trend, whereas the cognitive function is getting better over time. A rising trend is observed in the predicted number of chronic diseases. And a "V" shape is observed in positive feeling scores due to negative social and economic events in year 2008.

Chapter 9 investigates Chinese oldest old's smoking and alcohol use behaviors and their health status. Findings show that the smoking and drinking behaviors vary by time period and across birth cohorts. The percentage of the oldest old who smoked both in the past and at present show a declining trend over time. The percentage of constant smokers has been decreasing across birth cohorts, which suggests a healthier lifestyle among younger birth cohorts. Similar to smoking behavior, the percentage of the oldest old who constantly consumed alcohol also shows a declining trend. But no significant cohort changes are observed in such a percentage. This finding suggests that alcohol use behavior is relatively stable across birth cohorts.

The results drawn from the CCREMs also highlight the period and cohort effects when linking health disparities to the oldest old's smoking and drinking behaviors. Basically, when using smoking and drinking variables to predict various measures of the oldest old's health, SRH and negative feeling scores of the respondents show a declining trend over time. But the predicted number of chronic diseases and cognitive function scores demonstrate an increasing pattern by time period. The positive feeling score drops considerably in year 2008, indicating a significant period effect caused by destructive social events, such as economic recession and local natural disasters.

Significant cohort effects are shown in SRH, cognitive function and negative feeling scores. The results show that overall the cohort effects are not as apparent as period effects. There is no declining or increasing trend in SRH or negative feeling scores. But the cognitive function scores show a general declining trend with the 1901–03 birth cohort demonstrating the best predicted cognitive function scores than other birth cohorts. Crossovers are also shown in predicted cognitive function and negative feeling scores between subgroups, signifying that the health disparities due to smoking and drinking behaviors may shrink across birth cohorts.

Overall, the manuscript has shown significant period and cohort variations in Chinese oldest old's health outcomes as well as in the associations between the oldest old's health and a number of important health determinants. These facts would not have been discovered if the research of this book was not conducted. The findings have improved our understanding on the ever changing patterns of the oldest old's health. The results also help to improve existing demographic and sociological theories on the oldest old when a longitudinal perspective is brought into the scope of health research.

10.2 Implications

Findings of this book with regard to the period and cohort effects when studying Chinese oldest old's health contain important implications. These implications are vital not only for theoretical framework constructing but also for health policy making. In terms of the theoretical framework, it appears evident that existing gerontology theories based on cross-sectional analyses are not sufficiently satisfactory explanations for longitudinal variations in the oldest old's health. As this study shows, the oldest old's health is changing over time and across birth cohorts. Significant period and cohort variations are also found when using a variety of factors to predict the oldest old's health. Thus, the ability of conventional gerontology theories explaining the oldest old's health differentials is limited when taking the compound effects of period and cohort into consideration. For instance, when using several health determinants to predict the oldest old's health, chronic diseases show a curvy pattern instead of a linear trend across birth cohorts. The predicted positive feeling scores also show a "V" shape by time period. These results point out that merely relying on theories based on cross-sectional research findings to understand trajectories of the oldest old's health is not appropriate. Conventional gerontology theories based on cross-sectional analysis results are not able to accurately explain dynamic patterns of the oldest old's health over time.

Since health of the oldest old directly impacts the health care systems, disease prevention programs, and old age care burdens of families, communities and the society, findings of this research also have important policy implications. The results show that males, married individuals and urban elders generally have better health than females, unmarried and rural older adults. Socioeconomically more advanced groups also enjoy a better health condition in old ages than those with a lower SES status. These findings remind health professionals and policy makers that females, unmarried, rural and low SES elders should be main targets of the health care programs. They are in a more disadvantaged position and they need more care, support, resources and treatment in older ages.

A higher level of participation in leisure and physical activities is found to lead to a better health condition. Such a finding advocates that being physically and mentally active is crucial in maintaining a healthier status in old ages. If possible, caregivers should encourage the oldest old participating in a variety of physical and leisure activities to prevent health deterioration. Local government and communities should launch effective programs encouraging older adults to stay active. For instance, various employment opportunities can be offered to older adults. Some free or low cost traveling, gathering or cultural excursion programs can be made available to older adults as well to improve their social interaction with people and enlarge their social networks.

Findings of this research also point out significant period and cohort changes in the oldest old's health when health is predicted by a variety of demographic, socioeconomic, behavioral and lifestyle factors. Some of these findings are worth discussing. For instance, the predicted number of chronic diseases is rising by time period and across birth cohorts. This is an alarming sign that requires attention of health professionals and policy makers. China has completed the epidemiological transition from deaths of elders being caused primarily by transmitted diseases to chronic diseases. Chronic diseases now become main killers of older adults, especially among more recent birth cohorts. Health care policies should therefore pay more attention to chronic disease education, prevention and treatment.

The results also show that positive feeling scores of the oldest old are heavily influenced by social events, especially negative social events. For example, an economic hardship and natural disasters that occurred in year 2008 have left strong imprints on the oldest old's subjective wellbeing. These results suggest that as compared to physical health, the oldest old's mental health is more likely to be affected by negative social events. And the strong period effects are independent of the influence of other social and behavioral factors. These findings remind health professionals, caregivers, government officials and policy makers that we may need to pay a particular attention to the oldest old's mental health when negative social events occur. Educational programs may be launched to educate citizens, particularly elders when unexpected negative social events happen. Free or affordable psychological consultation should be made available to elders who have already developed depression symptoms.

Another interesting finding that needs to be discussed here is that the health advantage in cognitive function due to participating in leisure activities has diminished across birth cohorts. As it was documented in prior literature, engaging in a number of leisure activities is beneficial in maintaining older adults' cognitive function and preventing dementia (Bum et al., 2020; Hansdottir et al., 2022; Michele et al., 2019). The shrinking disparities in cognitive function due to leisure activity participation may suggest that we need to explore other forms of activities that may help the oldest old maintaining their cognitive function. Those types of activities can be internet and media use, moderate grandparental child care practice, social networking and so forth. Previous studies have repeatedly showed that all of these activities are beneficial in lowering older adults', particularly the oldest old's risks of cognitive function decline (Arpino & Bordone, 2014; Coget et al., 2002; Guo & Ling, 2022; Hayotte & Brunson, 2022; Silversteina & Zuo, 2021; Tang et al., 2020; Zhang et al., 2022a; Zhou et al., 2017). Health care providers and policy makers may consider bringing other forms of activities to elders' daily life to prevent their cognitive decline. Those activities are such as internet use, chatting groups, senior educational programs and so forth.

In this research, health disparities among the oldest old due to smoking and alcohol use are also found to be shrinking across birth cohorts. Together with the findings of leisure activities' declining effects across birth cohorts, does it mean that the influence of behavioral as well as lifestyle factors on the oldest old's health has been decreasing across birth cohorts? If so, what are the reasons that have caused this change? What are the other factors that have increasing impacts on the oldest old's health? Health researchers should probably consider the implications of the above findings and explore some other influential factors on the oldest old's health in more recent years and among more recent birth cohorts.

10.3 Future Prospects

This monograph is a pioneer study that has systematically examined the compound effects of age, period and cohort in the associations of a variety of demographic, socioeconomic, behavioral and healthy lifestyle factors and Chinese oldest old's health. The research has several limitations that need to be addressed. First, the study shows significant period and cohort effects, but it does not investigate why these effects exist. For example, the findings show that health of individuals with a lower SES is more sensitive to cohort and period changes as compared to health of the oldest old with higher a SES. Is this because the oldest old with a higher SES generally have more resources to deal with negative life events and get better medical treatments? What kind of social events and cohort changes have caused the period and cohort variations? These questions remain unanswered in this study.

In addition, when using health determinants to predict the oldest old's chronic diseases, a "V" shape instead of a rising trend is observed across birth cohorts. This finding is contradictory to what I have expected that chronic diseases should increase across birth cohorts. What are the reasons behind the "V" shape trend? Why does the middle birth cohort show a fewer number of chronic diseases than its earlier and later birth cohorts? This study obviously has not offered an explanation. Future research may extend this current analysis by further investigating the reasons that have caused the period and cohort variations when predicting the oldest old's health.

Further, when selecting health determinants that need to be included in the research, I have only focused on analyzing demographic, socioeconomic, behavioral and lifestyle factors. Since health is determined by a range of variables, there are a number of other important covariates that have not been examined in this research. Those factors are such as intergenerational relations, living arrangements, internet use, social networks, grandparental child care, religious participation et al. All of these factors have been shown to have significant influence on older adults'(including the oldest old's) health in prior literature (Bae, 2020; Chan et al., 2011; Chen et al., 2020; Cohen & Janicki-Deverts, 2009; Guo & Ling, 2022; Hayotte & Brunson, 2022; Headey et al., 2014; Idler, 2011; Li & Zhang, 2023; Noriega et al., 2022; Saha et al., 2022; Silversteina & Zuo, 2021; Wang, 2022; Yang et al., 2021; Zhang et al., 2019). Even under the lifestyle rubric, a number of important dimensions have not been covered in this study. These dimensions are such as sleeping, fruit and vegetable intakes, tea consumption et al., which have been found to have significant effects on older adults', especially the oldest old's health in the existing literature (An et al., 2019; Gehlich et al., 2020; Jyvakorpi et al., 2018; Ruan et al., 2013; Suzuki, Etsuji et al., 2009; Tao et al., 2023; Tsou, 2011; Zhao et al., 2021a). The above factors certainly deserve further analyses when examining the compound effects of age, period and cohort on the oldest old's health.

In addition, several measures used in this monograph are raw measures which can be refined in future research. Past studies have provided evidence that smoking and alcohol use have negative impacts on older adults' health (Alley et al., 2017; Barengo et al., 2019; Lahiri & Li, 2020; Lu et al., 2020; Sacco et al., 2015; Xiaohua et al., 2015; Zhang et al., 2020). Findings of this research, however, show that constant smokers and constant drinkers have better health outcomes than other subgroups. This result seems to be contradictory to previous findings. I urge readers interpret the result with caution. This is because the smoking and alcohol consumption variables used in this research cannot capture the levels of smoking and drinking behaviors. Thus, if the study merely relies on variables representing whether smoked or drank alcohol in the past, the results may well be biased. This is because those who smoked or consumed alcohol heavily in the past may have suffered from various types of diseases. Consequently, they quit smoking or drinking when they were surveyed. They show a worse health condition as compared to constant smokers and constant drinkers maybe because they smoked or drank heavily during the past. Due to the cumulative effects of smoking and drinking, they show worse health outcomes. For those who smoked and drank alcohol both in the past and at present may well be the group that have a stronger survival capability and a better health status even in childhood. They therefore demonstrate the best health outcomes in older ages even after they drank and smoked for a number of years. Thus, I recommend using variables that can capture levels of smoking and drinking to re-run the analyses to improve this current work.

As it was mentioned at the beginning of the monograph, low fertility together with a longer life expectancy have made aging related issues more challenging than ever for world countries. How to live longer and healthier at older ages becomes a key topic to study. This book has taken a longitudinal point of view to shed light on Chinese oldest old's health and its determinants and has revealed some important findings that have not been documented in previous studies. Future research may continue this work by bringing other social, behavioral, lifestyle and cultural factors into the analyses, such as intergenerational relations, living arrangements, internet use, religious participation, social networking and so forth to advance this current research. More importantly, future studies may further explore the reasons that have caused the period and cohort variations in the oldest old's health and investigate possible solutions to prevent detrimental influence of negative social events on the oldest old's health. More refined work is warranted to carry on this line of research on the oldest old's health.

References

Aihemaitijiang, S., Zhang, L., Ye, C., Halimulati, M., Huang, X., Wang, R., & Zhang, Z. (2022). Long-term high dietary diversity maintains good physical function in Chinese elderly: A cohort study based on CLHLS from 2011 to 2018. *Nutrients, 14*(9). https://doi.org/10.3390/nu14091730

Alley, S. J., Duncan, M. J., Schoeppe, S., Rebar, A. L., & Vandelanotte, C. (2017). 8-year trends in physical activity, nutrition, TV viewing time, smoking, alcohol and BMI: A comparison of younger and older Queensland adults. *PLoS One, 12*(3), e0172510. https://doi.org/10.1371/journal.pone.0172510

An, R., Liu, G., Khan, N., Yan, H., & Wang, Y. (2019). Dietary habits and cognitive impairment risk among oldest-old Chinese. *The Journals of Gerontology. Series B, Psychological Sciences and Social Sciences, 74*(3), 474–483. https://doi.org/10.1093/geronb/gbw170

Arpino, B., & Bordone, V. (2014). Does grandparenting pay off? The effect of child care on grandparents' cognitive functioning. *Journal of Marriage and Family, 76*, 337–351.

Bae, S. M. (2020). The association between health-related factors, physical and mental diseases, social activities, and cognitive function in elderly Koreans: A population-based cross-sectional study. *Psychogeriatrics, 20*(5), 654–662. https://doi.org/10.1111/psyg.12561

Barengo, N. C., Antikainen, R., Harald, K., & Jousilahti, P. (2019). Smoking and cancer, cardiovascular and total mortality among older adults: The Finrisk study. *Preventive Medicine Reports, 14*, 100875.

Bum, C.-H., Johnson, A., & Choi, C. (2020). Healthy aging and happiness in the Korean elderly based upon leisure activity type. *Iranian Journal of Public Health, 49*(3), 454–462. https://doi.org/10.18502/ijph.v49i3.3141

Chai, X., & Mei, J. (2022). Investigating food insecurity, health lifestyles, and self-rated health of older Canadians living alone. *BMC Public Health, 22*, 2264.

Chan, A., Malhotra, C., Malhotra, R., & Ostbye, T. (2011). Living arrangements, social networks and depressive symptoms among older men and women in Singapore. *International Journal of Geriatric Psychiatry, 26*(6), 630–639.

Chen, X., & Silverstein, M. (2000). Intergenerational social support and the psychological Well-being of older parents in China. *Research on Aging, 22*(1), 43–65.

Chen, G., Si, W., & Qiu, L. (2020). Intergenerational financial transfers and physical health of old people in rural China: Evidence from CHARLS data. *Ciência Rural, 50*(5). https://doi.org/10.1590/0103-8478cr20190725

Coget, J., Yamauchi, Y., & Suman, M. (2002). The internet, social networks, and loneliness. *IT & Society, 1*, 180–201.

Cohen, S., & Janicki-Deverts, D. (2009). Can we improve our physical health by altering our social networks? *Perspectives on Psychological Science, 4*, 375–378.

Gehlich, K. H., Beller, J., Lange-Asschenfeldt, B., Kocher, W., Meinke, M. C., & Lademann, J. (2020). Consumption of fruits and vegetables: Improved physical health, mental health, physical functioning and cognitive health in older adults from 11 European countries. *Aging & Mental Health, 24*(4), 634–641. https://doi.org/10.1080/13607863.2019.1571011

Guo, J., & Ling, W. (2022). The relationship between the mental health status and social support of the lonely elderly with government participation in the internet context. *Frontiers in Public Health, 10*, 1013069.

Guo, H., Lin, K., Yang, K., Ma, Z., Cao, M., Hu, Y., & Yan, Y. (2022). Trends of cancer incidence among Chinese older adults from 2005 to 2016: A log-linear regression and age-period-cohort analysis. *Frontiers in Public Health, 1023276.*

Hansdottir, H., Jonsdottir, M. K., Fisher, D. E., Eiriksdottir, G., Jonsson, P. V., & Gudnason, V. (2022). Creativity, leisure activities, social engagement and cognitive impairment: The AGES-Reykjavik study. *Aging Clinical and Experimental Research, 34*(5), 1027–1035. https://doi.org/10.1007/s40520-021-02036-1

Hayotte, P., & Brunson, L. (2022). Becoming a grandparent in the internet age: An integrative model of intergenerational relationships within the family and grandparents' satisfaction with life. *Journal of Intergenerational Relationships*. Published online.

Headey, B., Hoehne, G., & Wagner, G. G. (2014). Does religion make you healthier and longer lived? Evidence for Germany. *Social Indicators Research, 119*, 1335–1361.

Hou, C., Lin, Y., Ren, M., Liu, M., Ma, Y., & Li, H. (2018). Cognitive functioning transitions, health expectancies, and inequalities among elderly people in China: A nationwide longitudinal study. *Geriatric Psychiatry, 33*(12), 1635–1644.

Idler, E. L. (2011). *Religion and adult mortality: Group- and individual-level perspectives.* Springer.

Jyvakorpi, S. K., Urtamo, A., Pitkala, K. H., & Strandberg, T. E. (2018). Happiness of the oldest-old men is associated with fruit and vegetable intakes. *European Geriatric Medicine, 9*(5), 687–690. https://doi.org/10.1007/s41999-018-0084-9

Lahiri, K., & Li, X. (2020). Smoking behavior of older adults: A panel data analysis using HRS. *Journal of Quantitative Economics, 18*, 495–523.

Lee, Y., Kim, J., Back, J. H., Kim, S., & Ryu, M. (2013). Changes in combined lifestyle risks and disability transition in older adults: Korean longitudinal study of aging, 2006–2008. *Preventive Medicine, 56*, 124–129.

Li, T., & Zhang, Y. (2023). Intergenerational relationships across multiple children and older parents' depressive symptoms in China—A resource contingency perspective. *Journals of Gerontology: Social Sciences, 78*, 925–937.

Lu, Y., Sugawara, Y., Zhang, S., Tomata, Y., & Tsuji, I. (2020). Smoking cessation and incident dementia in elderly Japanese: The Ohsaki cohort 2006 study. *European Journal of Epidemiology, 35*, 851–860.

Luo, M., Ding, D., Bauman, A., Negin, J., & Phongsavan, P. (2020). Social engagement pattern, health behaviors and subjective Well-being of older adults: An international perspective using WHO-SAGE survey data. *BMC Public Health, 20*, 99.

Michele, J., Guillaume, M., Alain, T., Nathalie, B., Claude, F., & Kamel, G. (2019). Social and leisure activity profiles and Well-being among the older adults: A longitudinal study. *Aging & Mental Health, 23*(1), 77–83. https://doi.org/10.1080/13607863.2017.1394442

Noriega, C., Velasco, C., Pérez-Rojo, G., & López, J. (2022). Character strengths and social support as protective factors between grandparents' caregiving and health-related quality of life. *Journal of Child and Family Studies, 31*, 2505–2517.

Phulkerd, S., Gray, R., & Chamratrithirong, A. (2020). The influence of co-residential and non-co-residential living arrangements on sufficient fruit and vegetable consumption in the aging population in Thailand. *BMC Geriatrics, 20*, 476.

Ren, N., Li, Y., & Wan, Z. (2022). Patterns of HIV or AIDS mortality among older people from 1990 to 2019 in China: Age-period-cohort analysis. *JMIR Public Health and Surveillance, 8*(11), e35785.

Rizzuto, D., Orsini, N., Qiu, C., Wang, H. X., & Fratiglioni, L. (2019). Lifestyle, social factors, and survival after age 75: Population based study. *Journal of Aging and Health, 31*(7), 1297–1314.

Ruan, R., Feng, L., Li, J., Ng, T. P., & Zeng, Y. (2013). Tea consumption and mortality among oldest-old Chinese. *Journal of American Geriatric Society, 61*, 1–13.

Sacco, P., Unick, G. J., Kuerbis, A., Koru, A. G., & Moore, A. A. (2015). Alcohol-related diagnoses in hospital admissions for all causes among middle-aged and older adults: Trends and cohort differences from 1993 to 2010. *Journal of Aging and Health, 27*(8), 1358–1374.

Saha, A., Rahaman, M., Mandal, B., Biswas, S., & Govil, D. (2022). Rural urban differences in self-rated health among older adults: Examining the role of marital status and living arrangements. *BMC Public Health, 22*(1), 2175. https://doi.org/10.1186/s12889-022-14569-9

Silversteina, M., & Zuo, D. (2021). Grandparents caring for grandchildren in rural China: Consequences for emotional and cognitive health in later life. *Aging & Mental Health, 25*(11–12), 2042–2052.

Suzuki, E., Yorifuji, T., & Takao, S. (2009). Green tea consumption and mortality among Japanese elderly people: The prospective Shizuoka elderly cohort. *Annals of Epidemiology, 19*, 732–739.

Tang, D., Sun, H., & Xu, Y. (2020). The impacts of grandchild care on mental health among Chinese older adults: The mediating effects of social networks. *Population Research (in Chinese), 44*(4), 33–45.

Tao, L., Liao, J., Zheng, R., Zhang, X., & Shang, H. (2023). Association of drinking herbal tea with activities of daily living among elderly: A latent class analysis. *Nutrients, 15*(12). https://doi.org/10.3390/nu15122796

Tsou, M. T. (2011). Association between sleep duration and health outcome in elderly Taiwanese. *International Journal of Gerontology, 5*(4), 200–205. https://doi.org/10.1016/j.ijge.2011.09.020

Wang, F. (2022). Causal inference between the health status and living arrangements of elderly people in China. *Social Indicators Research, 159*, 1155–1178.

Wang, J., Liang, C., & Li, K. (2020). Impact of internet use on Eederly health: Empirical study based on Chinese General Social Survey (CGSS) data. *Healthcare, 8*, 482.

Wei, K., Liu, Y., Yang, J., Gu, N., Cao, X., Zhao, X., Jiang, L., & Li, C. (2022). Living arrangement modifies the associations of loneliness with adverse health outcomes in older adults: Evidence from the CLHLS. *BMC Geriatrics, 22*, 59.

Xiaohua, W., Yunqiang, W., Changxiang, C., Maoti, W., & Jie, Z. (2015). The effect of living habits such as smoking, alcohol drinking and green tea drinking on osteoporosis in the elderly. *Chinese Journal of Osteoporosis, 21*(10), 1187–1200. https://doi.org/10.3969/j.issn.1006-7108.2015.10.008

Xu, J., & Zhang, Q. (2023). The relationship between internet use and mental health of the elderly: Analysis of the differences between urban and rural. *PLoS One, 18*(1), e0280318. https://doi.org/10.1371/journal.pone.0280318

Yang, Y., & Land, K. C. (2006). A mixed models approach to the age-period-cohort analysis of repeated cross-section surveys, with an application to data on trends in verbal test scores. *Sociological Methodology, 36*(1), 75–97.

Yang, Y., Evandrou, M., & Vlachantoni, A. (2021). The impact of living arrangements and intergenerational support on the health status of older people in China. *Innovation. Aging, 5*(Supplement_1), 790.

Yang, Y., Xu, H., Liu, X., Li, J., Liew, Z., Liu, X., Huang, C., Zhu, J., Zhang, J., Chen, L., Hao, Y., Qin, G., & Yu, Y. (2022). Joint association of smoking and physical activity with mortality in elderly hypertensive patients: A Chinese population-based cohort study in 2007-2018. *Frontiers in Public Health, 10*, 1005260. https://doi.org/10.3389/fpubh.2022.1005260

Yu, Y. S., Ma, L., Lei, J., Li, J. F., & Ma, C. (2018). The relationship between community support and cognitive function in the elderly: Data analysis from the Chinese longitudinal healthy aging project. *Chinese Journal of Mental Health (in Chinese), 6*, 490–494.

Yuk, Y., Han, C. R., Jang, Y., Hong, Y. C., & Choi, Y. J. (2021). Association between weekly fruit and vegetable consumption and depressive symptoms: Results from the Korean elderly environmental panel study. *Epidemiol Health, 43*, e2021029. https://doi.org/10.4178/epih.e2021029

Zhang, L. (2021). A latent class analysis of health lifestyles and health outcomes among Chinese older adults. *Ageing & Society, 41*, 313–338.

Zhang, Y., Liu, Z., Zhang, L., Zhu, P., Wang, X., & Huang, Y. (2019). Association of living arrangements with depressive symptoms among older adults in China: A cross-sectional study. *BMC Public Health, 19*, 1017.

Zhang, R., Shen, L., Miles, T., Shen, Y., & Li, C. (2020). Association of low to moderate alcohol drinking with cognitive functions from middle to older age among us adults. *JAMA Network Open, 3*(6), e207922.

Zhang, J., Han, A., & Wu, X. (2021). Relation of depressive symptoms to intergenerational support and internet use in empty nest elderly. *Chinese Mental Health Journal, 35*(10), 838–843.

Zhang, Y., Su, D., Chen, Y., Tan, M., & Chen, X. (2022a). Effect of socioeconomic status on the physical and mental health of the elderly: The mediating effect of social participation. *BMC Public Health, 22*(1), 605. https://doi.org/10.1186/s12889-022-13062-7

Zhang, H., Wang, H., Yan, H., & Wang, X. (2022b). Impact of internet use on mental health among elderly individuals: A difference-in-differences study based on 2016–2018 CFPS data. *International Journal of Environmental Research and Public Health, 19*, 101.

Zhao, D., Ning, J., Zhao, Y., & Lu, E. (2021a). Associations of dietary and drinking water habits with number of natural teeth: A longitudinal study in the Chinese elderly population. *BMC Geriatrics, 21*(1), 525. https://doi.org/10.1186/s12877-021-02473-7

Zhao, Y., Song, J., Brytek-Matera, A., Zhang, H., & He, J. (2021b). The relationships between sleep and mental and physical health of Chinese elderly: Exploring the mediating roles of diet and physical activity. *Nutrients, 13*(4), 1316.

Zhou, J., Mao, W., Lee, Y., & Chi, I. (2017). The impact of caring for grandchildren on grandparents' physical health outcomes: The role of intergenerational support. *Research on Aging, 39*(5), 612–634.

Correction to: Living Longer and Healthier at Older Ages

Correction to:
Li Zhang, *Living Longer and Healthier at Older Ages*,
Advances in Studies of Aging and Health 3,
https://doi.org/10.1007/978-3-031-69773-9

The work on this publication was funded by the National Social Science Fund of China (Grant No. 21BRK043, grant recipient: Li Zhang), the Beijing Municipal Social Science Fund (Grant No. 23SRB008, grant recipient: Li Zhang), and the Distinguished Qian Duansheng Scholar Fund at China University of Political Science and Law.

Initially, this information was unfortunately not available and has now been included on the Copyright page as follows:

Funding
This research was funded by the National Social Science Fund of China (Grant No. 21BRK043, grant recipient: Li Zhang), the Social Science Fund of Beijing Municipal (Grant No. 23SRB008, grant recipient: Li Zhang), and the Distinguished Qian Duansheng Scholar Fund at China University of Political Science and Law. The funders had no role in study design, data collection and analysis, decision to publish, or preparation of the manuscript.

The updated version of this book can be found at
https://doi.org/10.1007/978-3-031-69773-9

L. Zhang, *Living Longer and Healthier at Older Ages*, Advances in Studies of Aging and Health 3, https://doi.org/10.1007/978-3-031-69773-9_11

Index

L. Zhang, *Living Longer and Healthier at Older Ages*, Advances in Studies of Aging and Health 3, https://doi.org/10.1007/978-3-031-69773-9

MIX
Papier aus verantwortungsvollen Quellen
Paper from responsible sources
FSC® C105338

If you have any concerns about our products,
you can contact us on
ProductSafety@springernature.com

In case Publisher is established outside the EU,
the EU authorized representative is:
Springer Nature Customer Service Center GmbH
Europaplatz 3, 69115 Heidelberg, Germany

Printed by Libri Plureos GmbH
in Hamburg, Germany